Critical Transitions
in Nature and Society

PRINCETON STUDIES IN COMPLEXITY

SERIES EDITORS
Simon A. Levin (Princeton University) and Steven H. Strogatz (Cornell University)

TITLES IN THE SERIES

MARTEN SCHEFFER

Critical Transitions in Nature and Society

Princeton University Press

Princeton and Oxford

Requests for permission to reproduce material from
this work should be sent to Permissions,
Princeton University Press

Published by Princeton University Press,
41 William Street, Princeton, New Jersey 08540
In the United Kingdom: Princeton University Press,
6 Oxford Street, Woodstock, Oxfordshire OX20 1TW

Library of Congress Cataloging-in-Publication Data
Scheffer, Marten.
Critical transitions in nature and society / Marten
Scheffer.
p. cm. —(Princeton studies in complexity)
Includes bibliographical references and index.
ISBN 978-0-691-12203-8 (hardcover : alk. paper)–
ISBN 978-0-691-12204-5 (pbk. : alk. paper)
1. Ecological disturbances.   2. Social evolution.
3. Threshold logic.   4. Global environmental
change.   5. Nature—Effect of human beings on.
6. Biogeography.   I. Title.
QH545.A1S34 2009
577.18—dc22
2009003871

British Library Cataloging-in-Publication
Data is available

This book has been composed in Minion

Printed on acid-free paper. ∞

press.princeton.edu

Printed in the United States of America

1 3 5 7 9 10 8 6 4 2

To Camila, Pablo, and Milena

.

Contents

Part II
CASE STUDIES

Part III
DEALING WITH CRITICAL TRANSITIONS

Acknowledgments

The idea to write this book came from Simon Levin and Sam Elworthy. Their support and that from Robert Kirk at Princeton University Press has been great. Also I am grateful to my colleagues at Wageningen University for being so flexible and supportive when it comes to my excursions into fields far beyond aquatic ecology.

This book greatly benefited from comments by Michael Benton, Frank Berendse, Victor Brovkin, Niels Daan, Toby Elmhirst, Carl Folke, Milena Holmgren, Terry Hughes, Johan van de Koppel, Tim Lenton, Andy Lotter, Jon Norberg, Max Rietkerk, Bob Steneck, and Brian Walker. Vasilis Dakos commented on the entire text, helped with the technical preparations, and also shared his emerging work on early warning signals.

The ideas in this book have been shaped over the years through my interaction with a few very special science friends. Sergio Rinaldi taught me how to think about dynamical systems theory in the graphical way laid out in this book. Buzz Holling invited me to his Resilience Alliance, a marvelous interdisciplinary group of people who have opened my eyes to a broad field of science that I might never have explored otherwise. Buz Brock and Frances Westley were my guides to the dazzling world of social sciences, and Steve Carpenter gave me wise advice in the turbulent world of science time after time. Last, Egbert van Nes has been my partner in science and stable point in space throughout the years.

Most of this text was written at the Chilean coast while enjoying the warm hospitality of Bjorn Holmgren and Ruth Urbá together with Pablo, Camila, and Milena, to whom the book is dedicated.

CHAPTER 1
Introduction

While our understanding of life at the molecular level is rapidly expanding, we understand remarkably little of the mechanisms that determine major transitions in societies or that regulate the stability and resilience of the climate and ecosystems such as coral reefs, forests, lakes, and oceans. Human population growth and development invokes a gradual but profound global change that has many faces. Atmospheric CO_2 levels rise steadily, concentrations of nutrients and toxic chemicals in lakes and rivers increase, groundwater levels drop, harvest rates from the ocean increase, and forests become increasingly fragmented. It is usually assumed that the response to such an incremental alteration of our planet will be gradual, predictable, and reversible. However, remarkably abrupt changes are occasionally observed in nature and society. Outbreaks of pests or pathogens and semiperiodic bursts of populations of rodents and other animals occasionally jump out of the blur. On a larger scale, the Earth system has evidently gone through rapid transitions between contrasting climatic conditions in the past, and it seems unlikely that similarly dramatic climate shifts would not happen in the future. And last, social systems are notorious for periods of relative inertia with occasional rapid transitions on scales varying from locally held opinions and attitudes to massive shifts such as the collapse of states and civilizations. In this book, I argue that such remarkable shifts can often be explained as so-called *critical transitions*. Just as a ship can become unstable if too much cargo is loaded on the deck, complex systems ranging from the climate to ecosystems and society can slowly lose resilience until even a minor perturbation can push them over a tipping point. While some critical transitions can play havoc on society, others represent escapes from undesirable situations. Understanding such transitions can open up surprising new ways of managing change. For instance, microcredit in the form of a small loan can allow a family to escape for good from the poverty trap, and a one-time intensive fishing effort can flip some lakes from a turbid condition to a stable clear state.

As an introduction to the theme of critical transitions, let us look at a few examples of surprising shifts. Although the causes of some of these shifts are quite well understood, others are still puzzling scientists.

1.1 Coral Reef Collapse

Caribbean coral reefs may be among the best-studied marine habitats. For decades, scientists of well-qualified research groups analyzed the structure and functioning of the reef communities. The reefs were thought to be resilient ecosystems. Incidentally, conspicuous changes, such as bleaching events and colonization of damaged sites by algae, were observed, but the reefs always recovered rapidly from such disturbances. Even when hurricane Allen caused extensive damage in 1980, the system recovered. There was a short-lived algal bloom, but within a few months, the algae disappeared and coral recruitment started filling up the open spaces created by the hurricane.

A few years later, however, a dramatic shift in the reefs took the research community by surprise and showed that it was time for a serious revision of the ideas about reef stability.[1] A species-specific pathogen caused mass mortality of the sea urchin *Diadema antillarum* with far-reaching consequences. The magnitude and impact of this event are well illustrated by the changes documented on Jamaican reefs (figure 1.1). Urchin densities crashed to 1% of their original level, and this triggered a complete change in the reef community. Rapidly the reefs became overgrown by brown fleshy algae that were now freed from grazing by urchins, and the entire community of the reefs changed profoundly. The apparent dependence of the system on

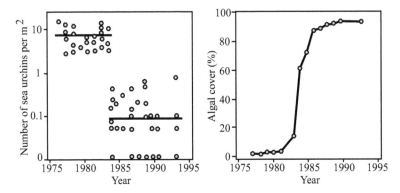

FIGURE 1.1. The Caribbean coral regime shift illustrated by changes on Jamaican reefs. (a) Collapse of populations of the sea urchin *Diadema antillarum* resulting from a pathogen outbreak. (b) Resulting increase in macro-algal cover at 7 m depth. (Modified from Hughes; see reference 1.)

a single species and the suddenness of the massive shift in the ecosystem were a surprise, but equally remarkable is the lack of recovery over most of the reefs until today.

1.2 The Birth of the Sahara Desert

A similarly striking *regime shift* seems to have happened on a very different timescale in the Sahel–Sahara region. It is difficult to imagine now that the Western Sahara has long been a relatively moist area with abundant vegetation and numerous wetlands. Yet this was the situation until about 6,000 years ago. Today, scattered bones of hippopotamus and other animals are reminders of this lush period. The dynamics of the change to the current state have been reconstructed by analyzing the amount of terrigenous dust in core samples of ocean sediments taken from the North African coast (figure 1.2). This sedimentary record suggests that for thousands of years the area was largely vegetated, until a remarkably swift transition to a desert state occurred.[2] Like the collapse of the coral reefs, this apparent desertification is a surprising deviation from what seemed to be the stable state of the system. Before this event, a mild trend existed, but certainly

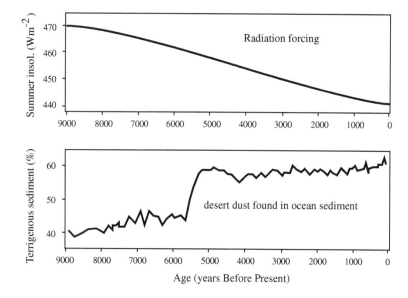

Figure 1.2. Over the last 9,000 years, average Northern Hemisphere summer insolation (top panel) has varied gradually because of subtle variation in the Earth's orbit. About 5,000 years ago, this change in solar radiation triggered an abrupt shift in climate and vegetation cover over the Sahara, as reflected in the contribution of terrigenous dust to oceanic sediment at a sample site near the African coast (bottom panel). (Modified from reference 2.)

extrapolating the dynamics over the millennia prior to 6,000 years ago would not suggest such a sudden and irreversible change. Imagine if we were living 6,000 years ago and had to predict the future of the region based on the information collected over the past 3,000 years. Certainly the best guess would have been a continuation of the trend that existed for thousands of years. How could we have foreseen the surprising shift? We now know that the climate change in the region is most likely related to subtle variations in the Earth's orbit, causing a gradual change in irradiation. However, this external variable changed very smoothly compared to the observed change in the area (figure 1.2). As you will see later, this ancient event is now explained by feedback between the vegetation and the climate system. But each time we unravel the mechanism behind such a transition in the past, the discovery immediately raises the question of whether similar events might happen again in the future and how we can predict or prevent that.

1.3 Shifts in Societies

Although this book is largely about shifts in nature, similar phenomena in human societies are too obvious to neglect. Stock market crashes are striking examples of unexpected sudden change. Also, the dynamics of public attitudes to certain problems is characterized by incidental massive shifts. Obviously, understanding such dynamics is important for politicians and anyone wanting to sell products or ideas. The book *The Tipping Point* by Malcolm Gladwell gives a persuasive account of social traps and shifts from that perspective.[3] Among the most dramatic societal shifts are collapses of nations into a state of anarchy and violence. From Rwanda and Somalia to Afghanistan and Bosnia, meltdown of governments has led to the displacement and deaths of millions of people. Having a clear image of what causes such state collapses and how they could be predicted would be invaluable. Indeed, the CIA has financed a multimillion dollar project to work on such questions.[4]

Looking further back in history, the dramatic demise of ancient cultures is one of the most persistent mysteries in the history of humankind.[5] It is puzzling that societies reaching such high levels of development disappeared so suddenly. Surely the people did not vanish altogether, but what remained was a society that left few archeological records.[6] Many explanations have been proposed for collapses of ancient societies, varying from depletion of vital resources to invasions of barbarians and catastrophes such as earthquakes, climate shifts, and floods. But this does not explain an interesting pattern that seems to underlie all of these cases. As Thor Heyerdahl put it, "The larger the pyramids and temples and statues they built in honor of their god or themselves, the harder was the fall."[7] Indeed, although adverse events may have triggered many of history's most well-known societal crashes, it appears as if complex, elaborate societies building impressive structures have been particularly prone to collapse. This suggests that the key to explaining ancient societal collapses may be found not only in external forces but also in the gradually increasing fragility of these elaborate societies.[8]

1.4 Content of this Book

As the rest of this book will show, these examples are just the tip of an iceberg. Reconstructions of shifts such as the ones in climate and societies that happened long ago are necessarily uncertain, and some may be even be artifacts of the way traces were preserved. In contrast, recent major changes that happen before our eyes are often quite well documented. Taken together, the evidence is overwhelming that sudden shifts to a contrasting state appear to occur from time to time in a wide range of systems. The term *regime shift* is often used to describe such radical changes.[9] Regime shifts seem often triggered by a major external impact. For instance, droughts, earthquakes, and floods hit ancient societies, and a disease outbreak led the Caribbean coral reefs to shift to an algal-dominated state. However, such disturbances may not always be the complete story. A much trickier aspect is that systems may gradually become increasingly fragile to the point that even a minor perturbation can trigger a drastic change toward another state. I will call such changes *critical transitions*. This book is about understanding, predicting, and avoiding or promoting such transitions in ecosystems, the climate, and societies. You might think it a bit sensational to address such overwhelmingly large phenomena. After all, radical turnover is the exception rather than the rule in most systems. However, there are two very good reasons why a focus on big, surprising dynamics makes sense: First, radical changes may be rare, but they are undeniably of crucial importance to society. Second, as you will see in this book, big eye-catching shifts are often driven by simple, identifiable dominant mechanisms. Therefore, these shifts are not only important, but also often more easily understood than the myriad of details.

This book consists of three parts. I start with an explanation of the basics of the theory of such forms of change. Subsequently, I review examples, and then I discuss how we might use our insight into the mechanisms of critical transitions from a practical point of view. This is what the three parts have to offer:

- **Part I** is a mini-guide to the theory of dynamical systems for people who have no affinity for math but who nonetheless enjoy a bit of abstract thinking. Dynamical systems theory is a powerful way of describing the essence of how any system, from a mix of chemicals to populations of organisms, the climate, or the solar system might behave. This theory is highly abstract, however. Rather than referring to any particular part of the world, it addresses what seems to be another world: a mathematical world of strange *attractors, catastrophe folds,* and *metastable states,* where *torus* destruction and *homoclinic bifurcations* are everyday events. So disparate is the language and notation in this discipline that it is difficult to imagine that it has any connection to reality as we know it. Nonetheless, underlying structures of the real world show up in the mirror world of math with a beautiful clarity that can never be seen in reality. This is not to say that the abstract models really explain anything. They simply hint at classes of mechanisms that may explain certain dynamics. Thus, dynamical systems theory goes only so far. The challenge is in pursuing these hints further to obtain a true understanding of forces that drive big changes in the real world. The final sections in Part I look at ways of bridging this gap and also give an overview of more informal, intuitive theories such as the idea that systems go through adaptive cycles of change. Although I am quite passionate about dynamical systems theory, I have tried to keep this introduction to the theory as basic as possible, and have even confined all the relevant equations to the appendix. Nonetheless, if you are easily overwhelmed by abstract concepts, you can jump directly to the case studies and read pieces of the theory chapter later as background.
- **Part II** consists of case studies and is a good place to start reading if your interests are more practical. This part gives examples of major transitions and oscillations in ecosystems, evolution, the climate, and societies, explaining little pieces of theory along the way. It becomes immediately obvious here that while essentially

similar phenomena occur across this broad range of complex systems, our understanding is much better for some phenomena than for others. I start this part with lakes, not because these are the most important systems, but because they can be relatively well understood and may serve as a model for more difficult systems such as the climate, the ocean, and societies. In fact, the classic article "The Lake as a Microcosm"[10] took this perspective, although it is now considered "merely" the start of the science of ecology. Evolution, the climate system, and human societies are at the other end of the range, in the sense that their dynamics are extremely difficult to understand and predict.

- **Part III** takes the practical perspective of how we can make good use of our insights in the phenomenon of critical transitions. How can we see transitions coming? How can we prevent bad transitions or promote good ones? Why are societies having difficulties in dealing with transitions, and how could we do better?

While the scope of this book is ambitious, I believe that the time is ripe to make headway in understanding the dynamics of the vast and complex dynamical systems that we live in. We have been good at unraveling how things work at a molecular level, but that is not enough. Few would deny that major changes are to be expected in ecosystems, climate, and society over the next century. We are in charge, but we need good science to help us shape the future in the best way—science that helps in understanding and managing the dynamics not only of chemical reaction vessels, but also of oceans, forests, climate, and society. As you will see, essentially the same kinds of mechanisms govern major transitions in such systems. However, it is striking how different the approach is between the branches of science. E. O. Wilson has argued that there should really be one approach to science covering everything from the atoms to the arts, and I would consider this book worth the effort if it contributes to the construction of such "consilience."

Part I
THEORY OF CRITICAL TRANSITIONS

This first part of this book is a simple guide to the parts of the theory of dynamical systems that are essential for understanding critical transitions. As will be the case throughout this book, I outline the key concepts by means of graphs. Examples with equations can be found in the appendix.

CHAPTER 2
Alternative Stable States

Suppose that you are in a canoe and gradually lean farther and farther over to one side to look at something interesting underwater. Leaning over too far may cause you to capsize and end up in an alternative stable state upside down. Although the details of the theory of alternative stable states may appear tricky, several key properties can be seen in this simple example. For instance, returning from the capsized state requires more than just leaning a bit less to the side. It is difficult to see the tipping point coming, as the position of the boat may change relatively little up until the critical point. Also, close to the tipping point, resilience of the upright position is small, and minor disturbances such as a small wave can tip the balance.

Resilience, defined as the ability of a system to recover to the original state upon a disturbance, is a tricky issue. The famous seventeenth-century Swedish ship the *Wasa* is an example in case. Named for the royal house of Wasa, the ship was built as the most prestigious ship of the navy of Gustavus Adolphus to combat the Polish on the Baltic. Against the advice of his engineers, the king insisted on an extra layer making the ship higher. The result looked impressive. However, when the ship set sail on its maiden voyage, a sudden breeze of wind made it sink within minutes, not a mile from land (figure 2.1). In a similar vein, we are reminded almost every year of the instability of ships by tragic accidents with ferries. Overloading is an especially treacherous aspect, as it can reduce resilience of the correct position in an

FIGURE 2.1. The resilience of the upright position of the seventeenth-century Swedish ship *Wasa* was apparently small enough to allow a sudden breeze to cause it to capsize in the first minutes of its maiden voyage.

unperceived way. Wave action or too many people moving to one side then tip the boat.

In such everyday examples of systems with alternative stable states, the consequences are intuitively straightforward. Nevertheless, with respect to complex systems such as societies and ecosystems, the idea is perhaps somewhat counterintuitive. Stability shifts are well known in mathematical theory, and the ideas expressed in an early book by René Thom about *catastrophe theory*[1] inspired a wide audience. Unfortunately, after this first inspiring work, the theory was swept up by the popular media, and claims were made that far outstripped its real capabilities. Just as chaos theory has been, catastrophe theory was quickly adopted by some almost as a philosophical doctrine or a general way of looking at things, without real, solid foundations. This caused some backlash among scientists against the applicability of the theory that set back progress. Still, much excellent work was done as a follow-up. In ecology, the field has been pushed forward in some pioneering papers in the 1960s and 1970s. Three articles have been particularly influential. First, in a systematic treatment, Richard Lewontin pointed out the theoretical possibility of alternative stable states.[2]

Subsequently, Crawford (Buzz) Holling linked these rather theoretical arguments to practical ecology in a more intuitive way,[3] and last, Robert May published a compelling review on thresholds and break-points in ecology that reached a wide audience.[4] Still, the work on such critical stability phenomena has remained largely theoretical for decades. Indeed, although the phenomenon of tipping points has many intuitive examples, it is less easily demonstrated that large and complex systems could also have them. Could coral reefs, the climate, or public attitude really tip over like a canoe? If so, can we manage or predict such shifts? As you will see, these themes are now again at the very forefront of scientific research, and I will spin them out in the rest of this book. However, we start with the basics.

2.1 The Basics

At the very heart of the theory of critical transitions are the notions of equilibrium and stability. In this section, I introduce those essential building blocks and show how tipping points can arise.

Equilibrium in Dynamical Systems

The theory of *dynamical systems* is a branch of mathematics used to describe the interaction of all kinds of processes. That sounds fairly broad, and indeed it is. A cell, a fish, a population, or the Earth may be considered dynamical systems—that is, their state can be understood as the result of a balance of processes. For instance, the temperature of the Earth is largely the result of gain of heat from solar radiation and loss of heat radiated back into the atmosphere. The result is a stable equilibrium. This means that if the temperature were brought into a slightly different state, it would move back to the original equilibrium. For instance, if the Earth were cooler than the equilibrium tempera-ture, it would radiate less into the atmosphere. Since the incoming radiation remains the same, this means that the Earth would warm up, thus moving in the direction of the equilibrium temperature. On the other hand, if the Earth were warmer, it would radiate more, and thus lose heat and cool down to the equilibrium again.

Another well-known example is that of a population that has reached the *carrying capacity* of the environment (figure 2.2). Such a population is at an equilibrium density resulting from a balance between birth and death rates. Again, it is a stable equilibrium. If a proportion of the population were wiped out by an adverse advent, there would be more resources for the survivors. This will tend to promote birth rates and reduce death rates so that the population will grow back to the equilibrium density. On the other hand, if densities exceed carrying capacity, reduced birth and increased mortality will push it back to the equilibrium. The overall rate of change in the population is the net result of gains and losses; in equilibrium, it is zero. In the appendix (see section A.1), you can find a classic equation for population growth and hints on how to play with this equation on a computer.

An attractive way to depict the stability of a system is by means of a stability landscape (figure 2.2, bottom panel). The slope at any point in such a landscape corresponds to the rate of change. Thus, at equilibrium, where the rate of change is zero, the slope is also zero. You can now imagine the system like a ball settling in the lowest point, representing the equilibrium. This "ball in a cup" analogue should not be taken too literally, as will be explained later. Nonetheless, it is a good intuitive aid for grasping the essence of the matter.

For obvious reasons, a *stable state* as depicted in these graphs is also called an *attractor*. The idea of an equilibrium is an important concept, but it is of course only the starting point for thinking about dynamical systems. The temperature of the Earth or the densities of populations are never constant. Fluctuations in the environment and all kinds of smaller or larger perturbations prevent equilibrium. Obviously, the equilibrium state can change if conditions are altered. For instance, as fossil fuel combustion causes greenhouse gas levels to increase, the Earth will retain relatively more heat and become warmer. Also, as you will see, there can be other kinds of attractors, such as cycles or strange attractors, that prevent systems from settling to a stable state, even if they were in a perfectly constant environment. Perhaps most importantly from the perspective of this book, systems can have several alternative attractors, separated by repelling points, cycles, or other structures. Critical transitions happen if a system shifts from one attractor to another.

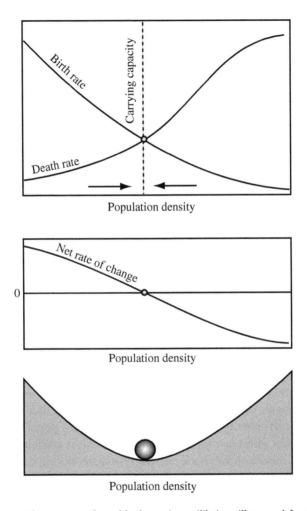

FIGURE 2.2. The concept of a stable dynamic equilibrium illustrated for the case of a hypothetical population that settles at a density that corresponds to the carrying capacity of the environment. The slope of the hills in the stability landscape (bottom panel) corresponds to the rate of change in population density (middle panel), which is the net result of per capita birth and death rates (top panel).

Alternative Equilibria

As a simple example of a system with alternative attractors, imagine a population of animals that runs into trouble if the numbers get too low. This may happen, for instance, if finding a mate becomes too difficult at low densities. Also, this can occur in species that congregate to protect themselves against predators. In such cases, mortality may go up at low population densities, but birth rates may fall too. For example, animals such as flamingos and penguins will not breed unless they are surrounded by many other mating individuals. The first to suggest that population growth can become depressed at low densities was the American zoologist Warder Allee, and the phenomenon is now commonly referred to as the *Allee effect*. Clearly, this is a highly important mechanism when it comes to understanding the extinction of endangered species. If the Allee effect is strong enough, it implies that a population can go into free fall if its density goes below a certain critical level (figure 2.3). In that case, the population has two alternative stable states: one at carrying capacity as in the case described earlier; the other at zero densities. In the previous simple growth model (figure 2.2), a zero density is also an equilibrium state, as even if per capita birth rates are high, absence of parents results simply in no offspring. In that case, however, a small addition of animals will be enough to kick off a population increase that stops only when the carrying capacity is reached. In contrast, a population with a strong Allee effect is trapped into the zero-state. A small initial population number is drawn back into the zero-state, as long as it is below the critical density (open dots in figure 2.3). As you will see later, plant populations can also have such a critical density. This happens especially under harsh conditions where a critical plant density is needed in order to "engineer" the environment sufficiently to make it suitable for plant growth. A simple mathematical model of a population with an Allee effect can be found in the appendix (section A.2).

You will see numerous other examples of systems with alternative equilibria throughout the book. However, before showing those examples, we will look a bit more closely at the general consequences for the dynamics of systems.

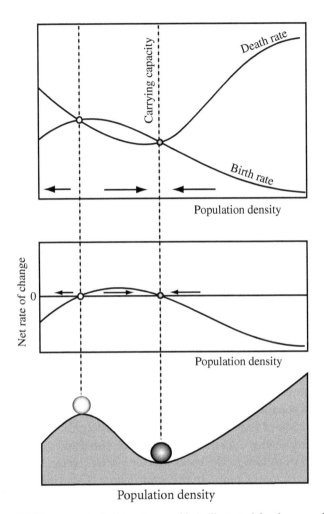

FIGURE 2.3. The concept of alternative equilibria illustrated for the case of a hypothetical population that experiences an *Allee effect*, implying that at low densities, the net growth of the population is negative because of increased per capita mortality and reduced per capita birth rates. As a result, a situation in which the population density is zero represents a stable state. Only if the critical density that marks the border of the basin of attraction of the nil state (small open circles in upper panels and light-colored ball in lower panel) is exceeded will the population grow and end up in the alternative stable state at carrying capacity.

CATASTROPHIC SHIFTS AND HYSTERESIS

The existence of multiple stable states has profound implications for the way in which a system responds to changing conditions. Mostly, the equilibrium of a dynamical system moves smoothly in response to changes in the environment (figure 2.4a). Also, it is quite common that the system is rather insensitive over certain ranges of the external conditions, while responding relatively strongly around some threshold condition (figure 2.4b). For instance, mortality of a species usually increases sharply around some critical concentration of a toxicant. In such a situation, a strong response happens when a threshold is passed. Such thresholds are obviously important to understand. However, a very different, much more extreme kind of threshold than this occurs if the system has alternative stable states. In that case, the curve that describes the response of the equilibrium to environmental conditions is typically "folded" (figure 2.4c). Note that such a *catastrophe fold* implies that indeed for a certain range of environmental conditions, the system has two *alternative stable states*, separated by an unstable equilibrium (dashed line) that marks the border between the basins of attraction of the alternative stable states, just as in the example of the population with an *Allee effect* (figure 2.3).

We can now see why this situation is the root of true critical transitions. When the system is in a state on the upper branch of the folded curve, it cannot pass to the lower branch smoothly. Instead, when conditions change sufficiently to pass the threshold (F_2), a "catastrophic" transition to the lower branch occurs (figure 2.5). Clearly, this is a very special point. In the exotic jargon of dynamical systems theory, it is called a *bifurcation point*. As you will see, there are several different kinds of bifurcation points that all mark thresholds at which the system's qualitative behavior changes. For instance, the system may start oscillating, or a species may go extinct at a bifurcation point.

The point shown in figure 2.5 marks a so-called *catastrophic bifurcation*. Such bifurcations are characterized by the fact that an infinitesimally small change in a control parameter (reflecting, for instance, the temperature) can invoke a large change in the state of the system if it crosses the bifurcation. Clearly, this kind of change is among the most interesting one from the perspective of this book. While all kinds

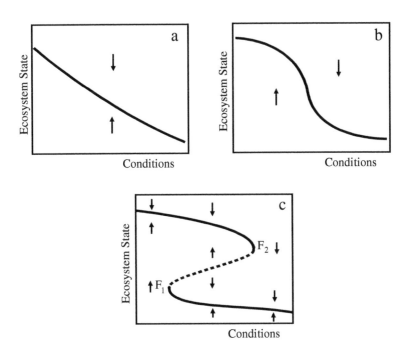

FIGURE 2.4. Schematic representation of possible ways in which the equilibrium state of a system can vary with conditions such as nutrient loading, exploitation, or temperature rise. In panels (a) and (b), only one equilibrium exists for each condition. However, if the equilibrium curve is folded backward as in panel (c), three equilibria can exist for a given condition. The arrows in the graphs indicate the direction in which the system moves if it is not in equilibrium (that is, not on the curve). It can be seen from these arrows that all curves represent stable equilibria, except for the dashed middle section in panel (c). If the system is pushed away a little bit from this part of the curve, it will move further away instead of returning. Hence, equilibria on this part of the curve are unstable and represent the border between the basins of attraction of the two alternative stable states on the upper and lower branches.

of bifurcations correspond in a sense to critical transitions, catastrophic bifurcations are really the mathematical analogue of the dramatic transitions we are trying to understand. The bifurcation points in a catastrophe fold (F_1 and F_2) are known as *fold bifurcations*. (They are also called *saddle-node* bifurcations, because in these points, a stable "node" equilibrium meets an unstable "saddle" equilibrium).

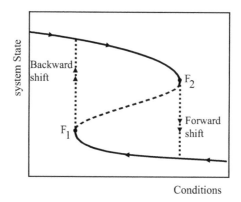

FIGURE 2.5. If a system has alternative stable states, critical transitions and hysteresis may occur. If the system is on the upper branch, but close to the bifurcation point F_2, a slight incremental change in conditions may bring it beyond the bifurcation and induce a critical transition (or *catastrophic shift*) to the lower alternative stable state (*forward shift*). If one tries to restore the state on the upper branch by means of reversing the conditions, the system shows hysteresis. A backward shift occurs only if conditions are reversed far enough to reach the other bifurcation point F_1.

The fact that a tiny change in conditions can cause a major shift is not the only aspect that sets systems with alternative attractors apart from the "normal" ones. Another important feature is the fact that in order to induce a switch back to the upper branch, it is not sufficient to restore the environmental conditions from before the collapse (F_2). Instead, one needs to go back further, beyond the other switch point (F_1), where the system recovers by shifting back to the upper branch. This pattern in which the *forward* and *backward switches* occur at different critical conditions (figure 2.5) is known as *hysteresis*. From a practical point of view, hysteresis is important, as it implies that this kind of catastrophic transition is not so easy to reverse.

The idea of catastrophic transitions and hysteresis can be nicely illustrated by stability landscapes. To illustrate how stability is affected by changes in conditions, we create stability landscapes for different values of the conditioning factor (figure 2.6). For conditions at which there is only one stable state, the landscape has only one valley, just as in the case discussed at the beginning of this section (figure 2.2). However, for the range of conditions where two alternative stable

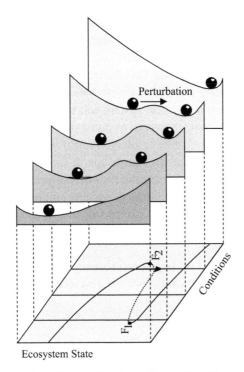

Ecosystem State

FIGURE 2.6. External conditions affect the resilience of multiple stable systems to perturbation. The bottom plane shows the equilibrium curve as in figure 2.5. The stability landscapes depict the equilibria and their basins of attraction at five different conditions. Stable equilibria correspond to valleys; the unstable middle section of the folded equilibrium curve corresponds to hilltops. If the size of the basin of attraction is small, resilience is small, and even a moderate perturbation may bring the system into the alternative basin of attraction.

states exist, the situation becomes more interesting. The stable states occur as valleys, separated by a hilltop. This hilltop is also an equilibrium (the slope of the landscape is zero). However, this equilibrium is unstable. It is a *repellor*. Even the slightest change away from it will lead to a self-propagating *runaway process* moving the system toward an attractor.

To comprehend the catastrophic transitions and hysteresis, imagine what happens if you start in the situation of the landscape in front. The system will then be in the only existing equilibrium. There is no other attractor, and therefore this state is said to be *globally stable*. Now suppose that conditions change gradually, so that the stability landscape changes to the second or third one in the row. Now there is an alternative attractor, implying that the state in which we were has become locally (rather than globally) stable. However, as long as no major perturbation occurs, the system will not move to this alternative attractor. In fact, if we were to monitor the state of the system, we would not see much change at all. Nothing would reveal the fundamental changes in the stability landscape. If conditions change even more, the *basin of attraction* around the equilibrium in which the system rests becomes very small (fourth stability landscape) and eventually disappears (last landscape), implying an inevitable catastrophic transition to the alternative state. Now if conditions are restored to previous levels, the system will not automatically shift back. Instead, it shows hysteresis. If no large perturbations occur, it will remain in the new state until the conditions are reversed beyond those of the second landscape.

Resilience as the Width of a Basin of Attraction

In reality, conditions are never constant. Accidental, or stochastic, events such as weather extremes, fires, or pest outbreaks can cause fluctuations in the conditioning factors but may also affect the state directly—for instance, by wiping out parts of populations. If there is only one basin of attraction, the system will settle back to essentially the same state after such events. However, if there are alternative stable states, a sufficiently severe perturbation may bring the system into the basin of attraction of another state. Obviously, the likelihood of this happening depends not only on the perturbation, but also on the size of the attraction basin. In terms of stability landscapes (figure 2.6), if the valley is small, a small perturbation may be enough to displace the ball far enough to push it over the hilltop, resulting in a shift to the alternative stable state. Following Holling,[5] I use the term *resilience* in this book to refer to the size of the valley or *basin of attraction* around a state that corresponds to the maximum perturbation that can be taken without causing a shift to an alternative stable state. A deeper discussion of different interpretations of the concept of resilience follows later in this book (see chapter 6).

A crucially important phenomenon in systems with multiple stable states is that gradually changing conditions may have little effect on the state of the system, but nevertheless reduce the size of the attraction basin (figure 2.6). This loss of resilience makes the system more fragile in the sense that it can be easily tipped into a contrasting state by stochastic events. This is also one of the most counterintuitive aspects. Whenever a large transition occurs, the cause is usually sought in events that might have caused it: The collapse of some ancient cultures may have been cause by droughts. An intractable conflict may be due to the act of an evil leader. A lake may have been pushed to a turbid state by a hurricane, and a meteor is thought to have wiped out the dinosaurs, leading to the rise of mammals. The idea that systems can become fragile in an invisible way because of gradual trends in climate, pollution, land cover, poverty, or exploitation pressure may seem counterintuitive. However, intuition can be a bad guide, and this is precisely where good and transparent systems theory can become useful. As you will see later, resilience can often be managed better than the occurrence of stochastic perturbations. Although such a resilience-based management style is starting to be used in some systems, it requires a major paradigm shift in many other areas. Part III of this book deals with practical questions, such as how insight into the mechanisms that determine resilience may help us manage systems with less effort to reduce the risk of unwanted transitions, and on the other hand promote good ones, such as escape from a poverty trap or the shift of a lake from turbid to clear.

The Continuum Between Catastrophic and Smooth Response

A tricky and often overlooked problem is that we can never generalize the stability properties of a system. For instance, we cannot make statements such as "The critical nutrient level for a lake to collapse into a turbid state is 0.1 mg phosphorus l^{-1}." In fact, we cannot even say that "Lakes have alternative stable states." In technical terms, the problem is that the position of critical bifurcation points (for example,

F_1 and F_2) always depends on various parameters of a model. In practice, this means that the corresponding thresholds are not fixed values. For instance, the critical nutrient level for a shallow lake to flip from a clear to a turbid state is lower for large lakes than for smaller ones.[6] In a wider sense, this means that safe limits to prevent critical transitions will usually not have universal fixed values.

A corollary is that the degree of hysteresis may vary strongly. For instance, shallow lakes can have a pronounced hysteresis in response to nutrient loading (figure 2.4c), whereas deeper lakes may react smoothly (figure 2.4b). Often a parameter can be found that can be changed such that the bifurcation points move closer together and eventually merge and disappear in so-called *co-dimension-2 points*. This particular point is known as the *cusp*,[7] after the hornlike shape produced by the fold bifurcations moving together (figure 2.7). It marks the change from a situation in which the system can respond in a catastrophic way to a situation in which the response to a control parameter (parameter 2 in figure 2.7) is always smooth because there are no alternative attractors. Thus, the panels with distinct possible responses to external conditions (figure 2.4) do in fact represent snapshots of a continuum of possible behavior that may be displayed by a single system. As you will see later, a positive feedback is usually responsible for causing a threshold response. A moderate feedback may turn a smooth response (figure 2.4a) into a threshold response (figure 2.4b), and a stronger feedback may cause the response curve to turn into a catastrophe fold (figure 2.4c).

Note also that there is in principle no limit to the number of alternative attractors in a system. In general, complex systems may have complex stability landscapes, with numerous smaller or larger attraction basins. In analogy to the scenarios for two alternative stable states, gradually changing conditions (for example, temperatures) may alter the landscapes, making some attraction basins larger and causing others to disappear. Meanwhile, disturbances occasionally flip the system out of an attraction basin that has become small, allowing it to settle into a more resilient state. Obviously, this is still a very idealized worldview, but before elaborating on it, we will review a few examples of mechanisms that may produce alternative stable states in models.

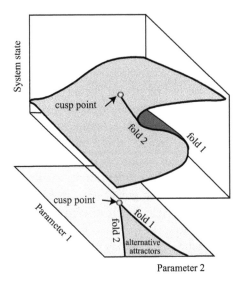

FIGURE 2.7. The cusp point where two fold bifurcation lines meet tangentially and disappear marks the change from a system with catastrophic state shifts in response to change in parameter 2 to one that responds smoothly to that parameter.

2.2 Some Mechanisms

Numerous mechanisms may lead to alternative stable states. The only way to get a good feel for how this phenomenon and the resulting dynamics can arise is to have a closer look at some examples. The following sections highlight a few. It may be challenging to really understand them; however, the reward will be to go beyond the superficial intuitive feel that you get from stability landscapes and tilting canoe stories. Equations and hints on how to analyze those on the computer are given in the appendix. More extensive discussions of case studies of regime shifts follow in part II of this book.

FACILITATION

The key ingredient to creating alternative stable states is a positive feedback that drives the system toward either of the states. If you start looking for it, such positive feedback is quite common in natural systems.[8]

In ecology, one of the most important mechanisms that can drive a positive feedback is *facilitation*. Although the attention of ecologists has focused largely on negative interactions such as competition and predation, there can also be conditions in which organisms have positive effects on the survival of others. There was considerable interest in such facilitation in the early work on succession in ecosystems. The idea was that pioneer species were essential to pave the way for later successional species. For decades, this topic has been out of fashion, but recently, there has been a strong revival of interest in facilitation. Experiments show that while competition is usually dominant under benign conditions, facilitation is common in harsh environments.[9] Although the phenomenon is not restricted to the plant kingdom, many examples have been documented for plants.[10] If the environment is harsh, facilitation effects may lead to alternative attractors if they are strong enough.

For instance, in dry environments, the microclimate may be ameliorated in the shade of large plants, where temperatures are lower and the humidity of soil and air is somewhat higher. If conditions are sufficiently arid, it is possible that seedlings may survive only under the canopies of such *nurse plants*. Obviously, this implies that in a completely barren situation, it is difficult to get vegetation started, even if the vegetated state can be stable due to the nursing effect of adult canopy.[11] The feedback of vegetation on moisture can also happen on much larger scales. In some regions such as the Sahel and part of the Amazon area, vegetation may promote precipitation. Loss of vegetation from such regions can potentially lead to a climate that is too dry to support the vegetation needed to keep up the precipitation (section 11.1).

Erosion prevention may be another way in which plants keep their environment suitable. Fertile soils can form with time under vegetation. Loss of vegetation may in some situations lead to soil erosion that makes vegetation recovery difficult. Similarly, on smaller scales, erosion of intertidal mudflats and lake sediments may be prevented by a layer of attached algae and other microorganisms. Initial consolidation can happen only in quiet periods, as normal resuspension and erosion prevents the organisms from establishing.[12]

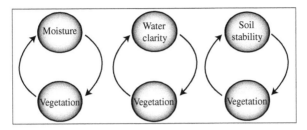

FIGURE 2.8. Three ways in which plants may have a positive effect on the environmental conditions that are essential for their growth.

A particularly well studied example of facilitation from aquatic environments is the development of submerged vegetation in turbid shallow lakes (see section 7.1). Submerged plants can greatly reduce turbidity by a suit of mechanisms such as control of excessive phytoplankton development and prevention of wave resuspension of sediments. However, the submerged plants also need low turbidity in order to get sufficient light. As a consequence, there can be situations in which loss of vegetation leads to an increase of turbidity sufficient to prevent recolonization by submerged plants.

In summary, there are various ways in which plants may have a positive effect on their own growth conditions (figure 2.8). It seems intuitively straightforward that this may lead to alternative stable states: one vegetated and another one without plants. However, things are more complex than that. First, alternative equilibria arise only if the feedback effect is strong enough (section 14.3). Second, stability of one of the states can be lost if external factors such as climate or nutrient input change (cf. figure 2.6).

To see how such loss of stability can occur, consider a simple graphical model of the response of shallow lakes to nutrient loading (figure 2.9). An overload with nutrients such as phosphorus and nitrogen derived from wastewater or fertilizer use tend to make lakes turbid. This is because the nutrients stimulate growth of microscopic phytoplankton that makes the water greenish and turbid. Although this *eutrophication* process can be gradual, shallow lakes tend to jump abruptly from the clear to the turbid state (see section 7.1 for more details). This behavior can be explained from a simple graphical model based

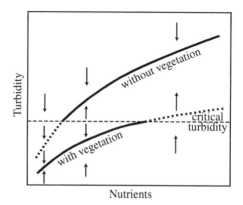

Figure 2.9. Alternative equilibrium turbidities caused by the disappearance of submerged vegetation when a critical turbidity is exceeded (see text for explanation). The arrows indicate the direction of change when the system is not in one of the two alternative stable states. (From reference 13.)

on only three assumptions: (1) turbidity increases with the nutrient level because of increased phytoplankton growth; (2) vegetation reduces turbidity; and (3) vegetation disappears when a critical turbidity is exceeded.

In view of the first two assumptions, equilibrium turbidity can be drawn as two different functions of the nutrient level: one for a macrophyte-dominated situation and one for an unvegetated situation. Above a critical turbidity, macrophytes will be absent, in which case the upper equilibrium line is the relevant one; below this turbidity, the lower equilibrium curve applies. The emerging picture shows that over a range of intermediate nutrient levels, two alternative equilibria exist: one with macrophytes and a more turbid one without vegetation. At lower nutrient levels, only the macrophyte-dominated equilibrium exists, whereas at the highest nutrient levels, there is only a vegetationless equilibrium.

The zigzag line formed by the stable and unstable equilibria in this graphical model corresponds to the folded line in figure 2.4c and the panel below the stability landscapes in figure 2.6. However, this simple example may serve to provide a better sense of the way in which a facilitation mechanism may cause the system to respond to environmental change showing hysteresis and catastrophic transitions. Gradual enrichment starting from low nutrient levels will cause the lake to proceed along the lower equilibrium curve until the critical turbidity is reached at which macrophytes disappear. Now a jump to a more turbid equilibrium at the upper part of the curve occurs. In order to restore the macrophyte-dominated state by means of nutrient management, the nutrient level must be lowered to a value where phytoplankton growth is limited enough by nutrients alone to reach the critical turbidity for macrophytes again. At the extremes of the range of nutrient levels over which alternative stable states exist, either of the equilibrium lines approaches the critical turbidity that represents the breakpoint of the system. This corresponds to a decrease of resilience. Near the edges, a small perturbation is enough to bring the system over the critical line and to cause a switch to the other equilibrium.

Analogous graphical models may be produced for other situations. For instance, imagine a dry region in which vegetation significantly improves microclimatic moisture conditions, as mentioned earlier. This may cause the system to respond in a hysteretic way to climatic change (figure 2.10). In analogy to the shallow lake model, the assumptions are: (1) microclimatic moisture increases with the overall climatic moisture conditions; (2) vegetation promotes microclimatic moisture; and (3) vegetation disappears below a critical microclimatic moisture level.

Obviously, these models are highly idealized. The most unrealistic assumption in such models is that vegetation would disappear completely at a critical condition (for example, turbidity or microclimatic moisture). In reality, there will always be some spatial heterogeneity, making some sites more suitable for plant growth than others. For instance, at shallow sites in a lake, submerged plants will be better able to capture light despite the turbidity of the water. Similarly, in a dry landscape, even if streams do not carry water, the valleys are usually moister, and therefore better for plant growth. The result of such spatial heterogeneity is that vegetation will disappear more readily from some sites than from others. This variation in the landscape, together with the strength of the feedback, largely determines whether there will

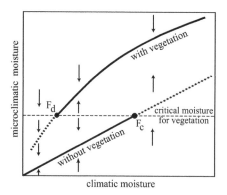

climatic moisture

FIGURE 2.10. A graphical model suggesting how alternative states with and without vegetation may arise in dry regions because of a positive effect of plants on microclimatic conditions.

be alternative equilibria and how large the hysteresis will be; this will be discussed further in the sections on case studies for lakes (section 7.1) and for terrestrial ecosystems (section 11.1).

Alternative equilibria may also arise from overexploitation of a natural resource. Overfishing and overgrazing are well-known examples. The biologist Noy-Meir[14] introduced an illuminating simple graphical approach to analyze the risk of overgrazing in dry rangelands. The idea is to plot the production of the food population and the losses due to consumption in the same graph (figure 2.11). The difference between the two can then be interpreted as the net growth of the food population.

The production of the food population plotted against its density shows an optimum. The section on logistic growth in the appendix (section A.1) explains the shape of this curve. Put simply, the explanation is that at low population densities, food-individuals grow and reproduce well, but since there are just a few individuals, overall productivity is still low. On the other hand, at the highest population densities, the productivity is low again because the carrying capacity of the environment is reached and severe competition prevents growth of the food population. The shape of the consumption curve can also be understood intuitively. Consumption drops to zero at low food density because the gathering of food by consumers becomes in-

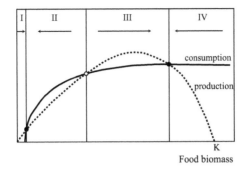

FIGURE 2.11. Graphical analysis of the stability of an exploited food population. At the intersections of the two curves, consumption equals production, and the food population is in equilibrium. The arrows indicate whether the food population biomass will increase or decrease in the different sections. From these arrows, it can be seen that the equilibrium on the intersection marked by an open circle is unstable, because the system will move away from it if one starts from a slightly higher or lower food population density.

creasingly difficult. For high food densities, it saturates at the maximum consumption rate of the consumer individuals. For a more detailed explanation and equations, see the appendix (section A.3).

The curves can intersect at three points (figure 2.11). All three intersections are equilibria, as consumption balances growth. Obviously, the food population will increase if production is higher than the losses due to consumption (sections I and III) and decrease if consumption exceeds production (sections II and IV). This implies that the middle intersection point is an unstable equilibrium. If the system is in this state (like a ball at the top of a hill), the slightest disturbance will cause it to move away to one of the stable states. This intersection represents the *breakpoint* below which the system collapses into an overexploited state with low food biomass where production is very low.

The essence of stability shifts can be understood from this graph if we follow what happens to the intersections as consumer density increases and decreases. Since the overall consumption increases with the amount of consumers present, the saturation level of the consumption curve will increase in proportion to the consumer density (figure 2.12). Starting from the lowest consumer density, there is just one

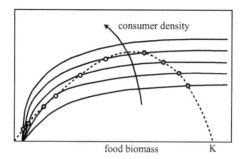

FIGURE 2.12. Consumption increases with consumer density, and this affects the position of the stable and unstable equilibrium points at the intersections of the production and the consumption curves.

equilibrium. With increasing consumer density, this equilibrium moves to the left. Food density decreases but productivity increases until the consumption curve becomes too high to intersect with the production curve. At that point, the equilibrium hits the unstable breakpoint and disappears. As a result, the system collapses into the overexploited state. If after this collapse consumer density is reduced in an attempt to restore the productive state, the system shows hysteresis. It stays in the overexploited equilibrium with low food densities until the consumption curve has become low enough to let the intersections at the left side disappear. Again this happens when the breakpoint collides with the stable state. Plotting the position of the three equilibria against consumer density (figure 2.13), one obtains a folded *catastrophe curve* that is analogous in interpretation to the ones discussed earlier (figure 2.6).

Not all types of exploitation lead to alternative equilibria. An important prerequisite is that the consumers are efficient in exploiting the food population even at low densities. In terms of figure 2.11, this implies that the consumption curve rises relatively steeply at low food densities. Only if consumption rises more steeply than food production on the lefthand side in the graph can the curves intersect at multiple points. Another prerequisite to obtain multiple intersections, and thus multiple stable states, is that the food population does not have too large an unexploitable fraction—that is, that the consumption curves starts rising at low food densities already. In the ap-

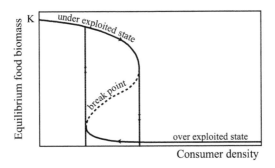

FIGURE 2.13. Food density in equilibrium plotted as a function of consumer density. The dashed middle section of the curve corresponds to the unstable breakpoints of the system (open circles in figure 2.12). In the range of consumer densities over which this unstable equilibrium exists, the system tends to either of the two alternative stable equilibria, depending on the initial density of the food population relative to the breakpoint.

pendix (section A.3), you can find a simple mathematical model of overexploitation.

THE POVERTY TRAP

In economics, various examples of positive feedback leading to alternative stable states have been identified. Perhaps most important to human society is the positive feedback in wealth that may cause a rich and a poor state to be alternative attractors.[15] The traditional economic view is that wealth is determined basically by effort. If you work hard enough, you become rich. Indeed, numerous success stories illustrate how famous people who started out poor achieved great wealth through hard work. Unfortunately, those belong to the same category as the stories of chain smokers reaching old age in good health. They are the exception to the rule. For instance, in the United States, famous for its trust in equal opportunities, a son born to parents in the poorest 10% of the population is 24 times more likely to stay in that poorest group as an adult than to end up as one of the 10% highest income earners.[16] A bimodal pattern of wealth is apparent not only for individuals, but also on various other scales. There are rich and poor countries, and this pattern has remained remarkably persistent over time. Similarly, within countries, there are often contrasting rich and poor groups, and within cities, there can be a distinction between persistently rich and poor neighborhoods. Traditional economic theory has difficulties explaining such patterns. Free trade should tend to equalize the wage rates of trading partners, and general *regression to the mean* should smooth unequal income distributions. But apparently this does not happen. More recently, new theories have emerged that may explain persistent poverty. They have addressed the puzzling existence of apparent *poverty traps* on various scales showing how poor economies may fail to develop altogether, why subgroups in rich economies may fail to share prosperity, and why a poor person may be unable to accumulate capital in a given society.[11] The overall image is that a number of different mechanisms may be involved. For instance, on the level of countries, corruption and organized crime may keep entire societies in an impoverished state, and for various other reasons, poor economies may simply be unable to produce the level of human and physical capital to achieve a certain kind of economic organization needed to escape the poor state. Within societies, positive feedback may maintain inequality in power and wealth. At the individual or family level, income may all be needed for food and shelter, allowing no investment in education or setting up a small business. Although the issue is complex, many of the models in this field lead to the prediction of a threshold of poverty that separates the poverty trap from an alternative richer condition (figure 2.14).

THE ICE–ALBEDO FEEDBACK

A well-known example of a positive feedback that acts on the scale of large regions or even the entire Earth system is the reflection of incoming radiation by snow and ice. As mentioned earlier, the temperature on the Earth depends on captured radiation from the sun and heat loss into space. If the albedo (the reflectivity) of the surface is high, more radiation is reflected and less is absorbed. Albedo can change strongly over time. For instance, if a region becomes covered by snow and ice, most of the solar radiation is reflected rather than captured. Therefore, the Earth becomes colder, and as a result, there will be a further increase in snow and ice cover. Of course, the flip

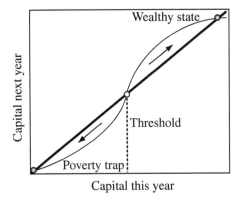

FIGURE 2.14. Simple graphical model of a poverty trap. The diagonal 45-deg line shows stable situations in which capital at time t equals capital at the next time-step. The sigmoidal line represents a model in which below a threshold capital level, the capital next year (at time $t + 1$) is smaller than the capital this year (at time t), implying a progression toward the stable poverty trap point. In contrast, if one starts above the threshold capital, growth toward the wealthy state will occur.

side of the coin is that warming results in less snow and ice, and since the darker Earth will then absorb more radiation, it will result in further warming. Models for predicting climate change, such as the ones used by the International Panel on Climate Change (IPCC), take this feedback into account, although large uncertainties exist. On a regional scale, the ice–albedo feedback may be particularly important. For instance, it is thought to amplify the warming of the Arctic, boosting the melting of the sea ice in the Arctic Sea over the coming century.[17] Also, in the past, the ice–albedo feedback has been a major amplifier of changes in temperature on the Earth. For instance, the fluctuations in temperature over the glacial cycles have probably been driven by subtle variations in radiation reaching the globe resulting from periodic variations in the orbit of the Earth around the sun. However, the major fluctuations in temperature and ice extent can be explained only if we account for the amplifying effects of the ice–albedo feedback and other internal mechanisms of the Earth system (section 8.2).

Mostly, the ice–albedo feedback is strong enough only to amplify a change and not to cause alternative stable states, but there may have been one spectacular exception to this rule in the past. One can imagine that the feedback might in principle lead to a runaway process.

More snow and ice would lead to lower temperatures, leading to more snow and ice, and so forth until the entire Earth is frozen. The Russian climate scientist Mikhail Budyko was the first to predict that this can indeed happen in theory.[18] His model predicted that the critical threshold for that would be reached if ice covered about half the Earth's surface area. Beyond that point, the feedback would cause a further shift toward a completely frozen planet. One of the reasons that the scientific community found this hard to believe is that it is difficult to imagine how the Earth could escape from such a trap to return to the current conditions. As explained later, evidence that our planet has indeed turned into a "snowball Earth" long ago (in *deep time*) has been found over the years, and a plausible escape mechanism has been suggested too (section 8.1).

2.3 Synthesis

In conclusion, alternative stable states can arise from a range of mechanisms. Usually, a positive feedback in the development toward one of the states can be identified as the basic mechanism. Whether alternative stable states will arise from a positive feedback depends among other things on the strength of the feedback. Obviously, the mechanisms shown here represent just a tiny subset meant as an illustration. Countless examples exist in chemistry and physics, ranging from the capsizing of ships to the onset of corrosion of stainless steel.[19] In ecology, alternative stable states can also arise from a range of other mechanisms in addition to the ones discussed here. For instance, an alternative stable state may commonly result from simple competition (see the appendix, section A.4), or if a predator shares a resource with its prey[20] or controls the natural enemies of its offspring,[21] or if populations live in scattered habitat fragments.[22] Similarly, runaway shifts are known from evolution[23] and various aspects of the climate system.[24] We will come back to some of those examples in part II of this book.

The key implication of the existence of alternative stable states is that gradual change in the environment can reduce resilience, allowing a small perturbation to cause a critical transition to a contrasting state. Such critical transitions are difficult to foresee and reverse.

CHAPTER 3

Cycles and Chaos

U p to now, we have looked at stable equilibrium points and unstable equilibrium points. However, these are only the simplest types of attractors and repellors. Given sufficient simulation time, many models of dynamical systems converge asymptotically to a cycle or a more complex dynamic regime rather than to a stable state. The theory on complex dynamics is extensive, and there is a risk of getting lost. However, from the point of view of critical transitions, it is important to have at least a basic idea of the different types of unstable regimes to which models may converge. There are two reasons for this. First, as explained in the next section, some systems may flip back and forth in an ongoing repetition of critical transitions because each of the alternate states has a slow destabilizing mechanism. Second, if internally generated cycles or chaotic fluctuations become too large, they may sometimes tip the system over the border of a basin of attraction, causing a larger shift to an alternative (stable, cyclic, or chaotic) attractor, just as external perturbations can do.

3.1 The Limit Cycle

The simplest alternative to the stable point is a so-called *limit cycle*. It is the root of much of complex dynamics and is therefore worthwhile understanding. The limit cycle implies that from different initial states, model runs converge to the same cyclic pattern. Many mechanisms may cause such behavior. Cycles may, for instance, result from time delays in regulation mechanisms. An example of the latter from economics is the occurrence of cycles in market prices and production of a certain goods. For instance, if the price of pork is high because of scarcity on the market, many farmers will start producing pork, resulting with time in an overshoot of production, causing prices to collapse and farmers to pull out of pork production. This results in pork scarcity and high prices, triggering the start of a new cycle.

Perhaps the best known example of a limit cycle is the one that can arise from the interaction between a consumer and its food population, most often referred to in the literature as the *predator–prey cycle*. In biological terms, what happens in this case is the following: The predator population eats almost all the prey. Because of the resulting food shortage, most predators die, and this allows the remaining prey individuals to reproduce and grow freely in an environment with few competitors and few predators. The resulting wealth of food for the few surviving predators, however, allows their population to expand too. The recovered predator population consumes almost all prey, and the whole cycle starts from the beginning again.

Most consumer–food interactions in nature do not generate such cycles, and there has been much interest in the question of what crucial conditions are needed to cause cycles as opposed to a stable equilibrium. A classic paper on this problem suggests that enrichment of the environment may cause a stable equilibrium to turn into a cyclic attractor.[1] As cycles bring the risk of extinction when population numbers become too low, this implies the surprising prediction that enrichment may lead to extinction, a finding referred to as the *paradox of enrichment*.

A classic mathematical model that produces predator–prey cycles is given in the appendix (section A.6). However, the background of this idea can be easily understood if we elaborate on the graphical model for overexploitation (figure 2.13) presented earlier. This model was built on the assumption that the consumer population density was a constant that could be manipulated. As a result, the system simply stabilized on either of the two stable parts of the catastrophe fold, depending on the initial condition (figure 3.1).

Obviously, in nature, consumer populations are not constant. In fact, they may often increase or decrease as a function of food avail-

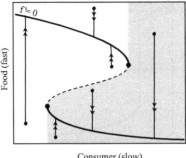

Consumer (slow)

FIGURE 3.1. Equilibrium biomass of a food population ($f' = 0$) as a function of the density of an efficient consumer (see figure 2.12). The consumer density is considered a constant that can be manipulated rather than a variable that responds to food availability. Starting from any point in the shaded area, the system will settle to the lower stable branch of the catastrophe fold, whereas from the unshaded area, it settles on the upper branch.

ability. The simplest assumption we can make about this is that there is a certain critical food density that is needed to maintain the consumer population. If the food density falls below this threshold, the consumer population declines. In contrast, if food abundance exceeds the critical level, the consumer population will increase. Thus, if we plot this critical food density in a graph, we have an equilibrium line for the consumer that separates the area of growth from that of decline of the population (figure 3.2).

CYCLES AS SELF-REPEATING CRITICAL TRANSITIONS

The interactive dynamics of the consumer and the food populations can be inferred from such a graph in a straightforward way if one assumes that the dynamics of one variable are much faster than those of the other variable.[2] This may seem artificial, but in fact rates of processes in ecosystems often differ widely. The difference in rates of change is an obvious problem if one must design field sampling programs meant to capture both fast and slow long-term changes. Also in simulation models, strong differences in the rates of processes are often considered a nuisance. Equations in which fast and slow processes are combined are known as *stiff equations*, which can create severe computational

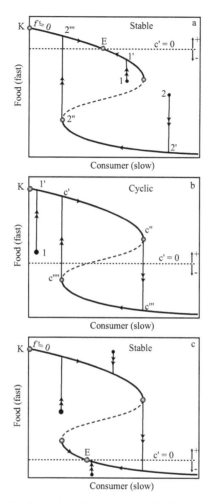

FIGURE 3.2. Interactive dynamics of a consumer and a food population, assuming the food dynamics to be much faster than the consumer dynamics. Note that often the consumer (or predator) is plotted on the vertical axis in such graphs.

problems. On the other hand, large differences in timescale can be used to our advantage. If we focus on the fast processes, we can simplify by neglecting the dynamics of the slow variables and consider them fixed. If we focus on the slow processes, we can assume that the fast ones react practically instantaneously to the slow changes, allowing us to replace fast dynamics with quasi-steady-state solutions. To analyze under which conditions our model gives rise to cycles, we follow a *slow–fast* approach, making the simplifying assumption that the difference in speed between consumers and food is very large.

In this situation, the trajectories of change are easy to draw. First, consider the example of the case in which the critical food density needed for consumer growth is relatively high (figure 3.2a). If one starts at point 1, the food population will quickly grow to equilibrium on the upper branch (1′) for the given consumer density. Since consumer dynamics are so slow, this fast piece of the trajectory is virtually vertical. From that point on (since there is enough food for the slow consumers to grow and the fast food population adjusts permanently to the equilibrium line), the system will slowly move along the food equilibrium line ($f' = 0$) to the stable equilibrium point E. This point represents the only attractor of the system, as trajectories from any starting point end up at E. Another example is the trajectory starting at point 2. It settles quickly to 2′, moves slowly to the left along the lower stable branch until it reaches the *bifurcation point* of the food population (2″), leading to a fast transition to the upper branch (2‴), from which it slowly settles to E.

The situation can become radically different if the critical food level for consumer growth ($c' = 0$) intersects the food catastrophe fold ($f' = 0$) in the unstable (dashed) part. In this case, there is no stable intersection point to which the system converges. Instead, the attractor to which all trajectories converge is a cycle (figure 3.2b). This is illustrated for one trajectory, starting at point 1 (figure 3.2b), but following the slow–fast approach, it can be easily seen that from all points, the system converges to the cycle c', c'', c''', c'''', c', and so on. This is a so-called slow–fast cycle, as it consists of slow movement along the food equilibrium curve (sections c' to c'' and c''' to c'''') punctuated by fast transitions between the stable branches. A typical example of slow–fast cycles in nature is the periodic outbreak of spruce budworm in boreal forests,[3] where the dynamics of the budworm are much faster than those of the spruce trees on which they forage (see also section 11.3).

A third possibility is that the critical food density ($c' = 0$) is low enough to allow an intersection with the lower ("overexploited") branch of the catastrophe fold (figure 3.2c). In that case, this intersection will be the only stable attractor of the system.

THE POINT OF DESTABILIZATION

We have seen that the system can be stable or cyclic depending on the shapes and intersections of the equilibrium curves of the two components. Different parameters may affect the lines, and hence the stability. For instance, nutrient enrichment may increase the carrying capacity for the food population (K). It can be easily seen from the graphs (figure 3.2) that if the critical food level ($c' = 0$) remains the same, an increase in K increasing the height of the catastrophe fold may change a system from stable (panel a) to unstable (panel b). This is the well-known paradox of enrichment.[1] The critical change occurs when the righthand extreme in the folded food equilibrium curve crosses the horizontal consumer equilibrium line. This changes the behavior of the system in a qualitative way. From a stable point, it becomes a cycle. This crucial point of destabilization is known as a *Hopf bifurcation*. As with any bifurcation, one may consider it a critical transition. However, it is not a catastrophic one. The initial change around the bifurcation is small, and there is no hysteresis. In the extreme case of the slow–fast system we analyzed, the cycle immediately takes it full size. However, in systems in which the difference in timescale between the two variables is less extreme, the cycle usually begins small and then grows to a larger size (figure 3.3). Note that the equilibrium point still exists after the system becomes destabilized. It has just become the unstable focus of the cycle, corresponding to the dashed middle section of the folded food equilibrium curve (figure 3.2). A simple mathematical model showing a Hopf bifurcation can be found in the appendix (section A.7).

3.2 Complex Dynamics

Although limit cycles do seem to appear in nature and society, usually dynamics are much more complex. Environmental fluctuations pushing the system around are an obvious cause of complex dynamics.

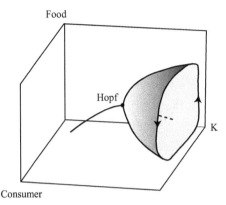

FIGURE 3.3. Schematic representation of a Hopf bifurcation marking the change from a stable equilibrium to a cyclic attractor evolving around the equilibrium point that has now become unstable. This particular example illustrates how enrichment (increase in the carrying capacity K) can destabilize a consumer–food system, a phenomenon known as the *paradox of enrichment*. This is called a "paradox" because we usually think of enrichment as a good thing.

However, dynamics generated by intrinsic mechanisms can be complex too. A major cause of complex dynamics is the interaction between two or more cycles. The simplest starting point to see this is to look at what happens if a cycling system, such as a predator–prey couple is influenced by a periodic oscillation of the environment. Such *periodic forcing* of biological systems by diurnal, seasonal, or tidal environmental cycles is virtually omnipresent. A mathematical model showing the effect of periodic forcing for the case of plankton can be found in the appendix (section A.10).

QUASI-PERIODIC OSCILLATIONS

If the biological system simply has a stable equilibrium point and is fast enough to respond, periodic forcing will merely turn the equilibrium into a cycle. (Similarly, a system with two alternative stable states separated by an unstable repelling point will turn into a system with two alternative stable cycles separated by an unstable repelling cycle.) However, if the biological system itself has an intrinsic cycle, the effect of periodic forcing is more complex. The simplest possibility is that

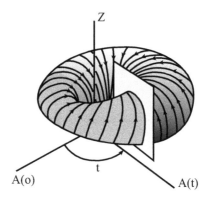

FIGURE 3.4. Mild periodic forcing of a system (in this case of algae A and zoo-plankton Z) can result in an attractor known as a *torus* if the nonforced system already had a limit cycle. Time (t) is represented in the graph by the angle between the axis A (0) and a rotating axis A (t), which returns to A (0) in one year so that one period of forcing corresponds to a turn of 360 deg around the central vertical axis. The depicted transverse frame is a so-called *Poincaré section* representing annual samples of the amount of algae and zooplankton on a fixed day of the year. (Adjusted from reference 4.)

the biological cycle is slightly and smoothly deformed. For instance, fast plankton cycles may become smaller if the temperature is lower and larger if the temperature is higher, so that a mild seasonal cycle in temperature results in a gradual swelling and shrinking of the cycle. Such a deformation leads to an attractor known as a *torus* (figure 3.4). A simulation run follows a trajectory that starts at an initial point on a torus and develops entirely on this doughnut-like structure but never comes back to the initial point, thus covering the torus densely as time goes on. Put otherwise, if one were to go out to sample this hypothetical plankton system each year on, say, the first of May, one would never find the same abundance as recorded in another year. Because of the lack of real periodicity, such an attractor is also referred to as a *quasi-periodic regime*.

PHASE LOCKING

If the periodic forcing is stronger, more interesting things start to happen. The forced cycle (for example, the plankton cycle) may start to lose its independence and start to dance to the beat of the forcing cycle. Such *frequency locking,* or *phase locking,* is a common phenomenon in nature. What will happen precisely depends on the ratio between the frequency of the forcing cycle and that of the forced cycle. If the frequencies are almost equal, the forced cycle will be easily locked to the forcing cycle. For instance, a human menstrual cycle is intrinsically close to a month and, in a significant portion of women, is locked to be in synchrony with the moon cycle.[5] If the frequencies are more dissimilar, locking becomes less likely. On the other hand, if the frequency of the forced cycle is close to two times the frequency of the forcing cycle, it will tend to lock in such a way that the biological cycle beats two times in each forcing round. Similarly, it can lock to a 1:3 rhythm or a 1:4 rhythm. In fact, there is no end to the possibilities. Locking may occur at 2:3, 3:4, 2:5, 1:6, 3:7, 4:9, and so on. However, not all those possibilities occur equally easily. Systems are more likely to lock to the simpler frequency ratios such as 1:2 than to the more complex ones such as 4:9. Whether locking will occur depends not only on the ratio of the frequencies but also on the strength of the forcing. Thus, if the frequency of the forced cycle is already very close to that of the forcing function, little forcing strength is enough to cause locking. However, a dissimilar frequency may still be locked if the forcing strength is sufficient to a certain limit. One way to capture all of this is to plot where locking occurs as a function of frequency and forcing strength (figure 3.5).

The fact that with sufficient force (amplitude) of the forcing function, even relatively dissimilar frequencies can be locked to, say, a 1:1 or a 1:2 ratio has an interesting consequence. Different locking ratios may represent alternative attractors. Thus, a forced cycle may be locked, for instance, to the same forcing cycle in a 2:3 or a 1:2 ratio. Once it is locked to either of those, it will tend to hang on to it, even if the frequency of the forcing function is gradually increased. At some point, it will switch to a different locking ratio, and if the forcing frequency is then gradually reduced, the system will show hysteresis in the sense that it tends to remain locked to the same ratio beyond the earlier switching point. Violinists or players of wind instruments may recognize the tendency of their oscillators (instruments) to hang on to an unwanted flageolet (for example, an octave, that is, a double frequency).

Phase locking is common in nature. For instance, the heart is really like a big orchestra with a director that governs the beat. The majority of

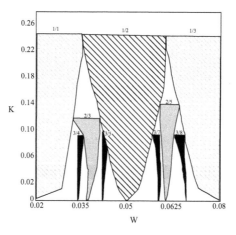

FIGURE 3.5. Windows of phase locking in a seasonally forced cyclic predator–prey model. The shaded areas represent ranges where the system is locked to oscillate in a fixed frequency related in a simple way to that of the forcing function (1/1, 1/2, 1/3, 2/3, 2/5, etc.). Along the horizontal axis, the frequency (W) of the forcing function is varied, while the strength of forcing (K) is plotted along the vertical axis. It can be seen that if the forcing is stronger, it can lock the frequency of the forced system over a larger range. At strong forcing, the areas with different frequency ratios overlap (not shown) as alternative attractors. (From reference 6.)

the heart cells are not spontaneously active. Instead, they are *excitable*— that is, they need to be driven into activity by some stimulus. The appropriate stimuli are produced in a specialized pacemaker region of the heart that does contain spontaneously active cells. Normally, the intrinsically quiescent cells respond in a 1:1 rhythm to the pacemaker beat. However, things can go out of sync when the excitability of the cells decreases, when there are problems in the conduction of electrical activity from cell to cell, or when the heart rate is raised too much. Looking into the response of an isolated cell to a pulsed stimulus, it can be seen experimentally as well as in simulation models that phase locking tends to occur in the sense that the synchronization shifts to a 2:1, a 3:2, or another ratio. Also, hysteresis occurs in the sense that the frequency of the beat needed to let the excitable cell flip into a new rhythm is different from the forcing frequency needed to flip it back.[7]

Numerous other cycles can be locked into phase, ranging from trees with synchronously blinking fireflies, to sleep–wake rhythms (locked to diurnal light–dark oscillation), to human menstruation cycles (locked to moon cycle or to a common cycle within groups of women). The climate system is also full of periodically forced cycles. For instance, as explained in detail later (section 8.2), glaciation cycles are forced by periodic variation in the Earth's orbit, and climatic oscillators such as El Niño Southern Oscillation (ENSO), the North Atlantic Oscillation (NAO), and the Pacific Decadal Oscillation (PDO) cycles are forced by the seasons.

In summary, periodic forcing of cyclic systems can lead to a range of dynamics. Quasi-periodic behavior on a torus is most common if the forcing strength is small. Stronger forcing usually results in phase locking, but as shown in the next section, chaotic dynamics may also arise. Alternative attractors with different frequencies can also result from periodic forcing of a cyclic system.

CHAOS

No doubt the most famous type of complex dynamics to which models may converge is *chaos*. This occurs frequently in periodically forced models but arises also from other mechanisms. The corresponding attractor is called a *strange attractor*, and for good reasons. In three dimensions, strange attractors often appear as dazzlingly elegant structures (figure 3.6). Even if they look relatively simple from a distance, they hide a surprising amount of detail. Like a quasi-periodic system, a chaotic system tends to a state of continuous change in which the same pattern is never exactly repeated. However, an important special feature of a chaotic system is that small differences in initial state expand exponentially with time. The sensitivity to initial conditions implies that the long-term behavior of a chaotic system is fundamentally unpredictable. Even if we know exactly the rules that govern the system, the final result remains unpredictable, because we can never precisely determine the current state. In fact, even if we could, the slightest perturbation has huge effects on the long term.

The weather is a well-known example of a chaotic system. Indeed, the first description of the phenomenon that became widely known comes from the study by Lorenz of an early computer model of atmospheric circulation. He found that his model never settled to a stable state or even a periodic motion and hit upon the sensitivity to initial

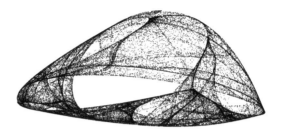

FIGURE 3.6. Example of a strange attractor produced by dots plotted at regular time intervals during a long simulation of the dynamics of a chaotic system.

conditions by accident.[8] Lorenz wanted to repeat a stretch of simulation he had done and retyped the values of the variables he read from the output as a starting condition. To his surprise, the simulated weather evolution started out the same but soon diverged to develop into something completely different. It appeared that the difference in the last decimals that had not appeared on the output he reentered was expanding exponentially in time in the simulations.

Clearly, this kind of chaos is not just what is meant by the word *chaos* in general. It is therefore more precise to refer to this as *deterministic chaos*. This term is used to stress that this unpredictable behavior can result from completely deterministic rules. Although the words *chaos* and *unpredictable* may sound rather discomforting, some qualifications to the unpredictability of chaos should be kept in mind. First, the strange attractor is always bounded. One can draw a box around it from which it will not escape. This implies that although the precise dynamics cannot be predicted, the range of behavior can be described, and it can be foreseen that this range will remain constant as long as external conditions do not change. For instance, although the weather in The Netherlands cannot be predicted well more than a few days ahead, it can be foreseen that it will not change into the weather of the Bahamas (as long as we do not alter the Earth system too much). Also, in chaotic attractors such as those found in competition systems, the sum of the variables shows relatively little variance over time and is therefore quite predictable. For instance, models of numerous competing algae may show chaotic fluctuations in the abundance of all the species, but total algal biomass remains relatively constant.[9] Last, although long-term predictions are difficult, there are several ways to predict short-term behavior. One approach is to model the mechanisms and simulate the behavior, as is done for most weather predictions. However, one may also study the attractor and make empirical models of its behavior. This can be done, for instance, by making a library of many short pieces of trajectory on the attractor (many fragments of the time-series). A way to predict the future is to then search the fragment that most resembles the recent past and use what happened after that fragment as a prediction of what will happen again in the near future.

Many mechanisms may produce chaos. In fact, a single difference equation of population growth (that is, computing the value of a variable at time $t + 1$ from that at time t with a fixed, relatively large time-step) may already lead to chaos.[10] Such a difference equation for computing stepwise changes in population densities may approximate, for instance, the case of insects with discrete generation cycles. The mechanism causing cycles and chaos is essentially a delay in the response to overpopulation. For instance, a generation living in a time of abundant food may produce many eggs, which results in an abundant next generation that overshoots the carrying capacity. Severe competition leads to poor reproduction, and the next generation may live in a time of abundant resources again, setting off a new cycle. If the growth rate is increased, the overshoots become wilder, and the system moves from stable through cyclic to chaotic.

If one simulates smooth change over time (using differential equations), three equations is the minimum needed to produce chaos in a constant environment. For instance, in ecology, chaotic dynamics can arise from a consumer exploiting two competing prey species, two consumers exploiting two separate but competing prey species, and a predator on top of a simple consumer–food system (figure 3.7). Although these interactions are potentially seeds of chaos, models for each of them produce chaotic behavior only for a restricted range of parameter settings. Therefore, the key question is not which kinds of systems might produce chaos, but rather when chaos will arise. Going into the details of this for any specific model appears not to be very illuminating, but there is an important generic rule: systems that contain interacting oscillators can easily become chaotic.[11] Since

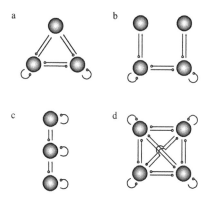

Figure 3.7. Some interaction structures for which simple models have been demonstrated to produce chaotic behavior. Arrows indicate positive effects, and lines with dots instead of arrowheads indicate negative effects. (a) A consumer exploiting two competing prey species, (b) two consumers exploiting prey that compete, (c) a carnivore on top of a simple consumer–food system, and (d) a network of four competing species. (From reference 14.)

predator–prey systems have a tendency to oscillate, this explains why combinations of linked predator–prey systems easily behave chaotically. However, they are certainly not the only source of chaos in ecology. For instance, a simple network of competing species can also behave chaotically.[9,12,13] Further, as mentioned earlier, periodic forcing and delayed feedback can induce chaos. As explained a bit later in part II of this book, such mechanisms may be important in driving irregularity in climatic phenomena such as El Niño and glaciation cycles.

3.3 Basin Boundary Collision

One reason cycles and chaos are important from the perspective of critical transitions is the fact that they can tip the system into the attraction basin of another attractor. Clearly, the fact that attractors can often be cycles, strange attractors, or other dynamic structures rather than simple stable points implies that we have to take a slightly different look at our intuitive stability landscape model of catastrophic shifts (figure 2.6). Apparently, rather than rolling to the bottom of the stability basin, balls may move around perpetually. This concept is hard to capture in such a graph, but a major implication may still be grasped from the stability landscape analogue. The dynamics of the wobbling system on its unstable attractor may cause it to hit the border of the attraction basin, invoking a shift to an alternative attractor. In a sense, this effect of internally generated turmoil causing a *basin boundary collision* is quite analogous to the effect of an external perturbation forcing the system into an alternative basin of attraction.

When Cycles Hit the Border of a Basin of Attraction

To grasp the essence of a basin boundary collision in terms of the theory of alternative attractors, we take a concrete example: fish feeding on zooplankton, which feeds on phytoplankton. In lakes, the key functional group of zooplankton is that of large herbivorous grazers. They are the only ones that can consume most size classes of phytoplankton and effectively clear the water. Since this type of zooplankton is easily spotted by small fish, they can be almost driven to extinction if sufficient fish are present. This corresponds to a classic overexploitation collapse (section 7.2) of zooplankton when a critical fish density is exceeded. There is an interesting twist to this story in the three-level food chain, as the zooplankton collapse releases phytoplankton from topdown control. Thus, at a critical fish density, we get a shift to an alternative state at two levels in the food chain, resulting in a green-water state with high algal biomass and little zooplankton (figure 3.8a).

Now, imagine what happens if in the absence of fish, zooplankton and phytoplankton settle not to an equilibrium, but rather to an oscillating regime.[15] In this case, if one slowly increases the fish density, the cycle comes increasingly close to the limit of the attraction basin of the alternative green state (figure 3.8). Eventually, when a critical fish density is reached, the cycle hits the limit of the attraction basin. If fish density is increased a little bit further, the cycle ends up in the attraction basin of the alternative stable green equilibrium state (figure 3.8). A mathematical model showing this phenomenon for a plankton system can be found in the appendix (section A.9). The situation in which the cycle precisely hits the boundary of the attraction basin is a type of bifurcation known as a *homoclinic bifurcation*. But the term *basin boundary collision*[16] is more intuitive. It belongs to a different family of bifurcations than the Hopf, the fold, and the trans-critical

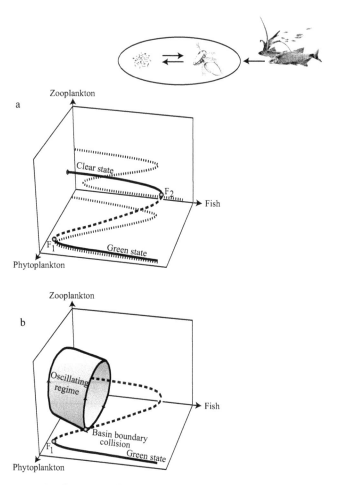

FIGURE 3.8. Simple catastrophic transitions (a) versus basin boundary collision (b) illustrated by a model of the effect of fish on plankton. Panel (a): If the plankton attractors are not oscillatory, a classic catastrophe fold arises where an increase in fish beyond the bifurcation point F_2 would lead to a collapse of zooplankton into an overexploited *green state* in which phytoplankton is abundant because of a lack of grazing. Return to the clear state at which zooplankton controls phytoplankton happens if fish are reduced to below the other fold bifurcation at F_1. Panel (b): If zooplankton and phytoplankton are involved in an oscillating attractor, the shift to the green state can occur through a basin boundary collision, also known as *homoclinic bifurcation*. Note that the oscillating regime and the green state are alternative attractors over a range of fish densities.

bifurcations we have looked at earlier. While those were *local bifurcations* that all had to do with stability changes at equilibrium points, the basin boundary collision is a nonlocal bifurcation that happens somewhere on a cycle or, as you will see later, on a more complex attractor. This may not seem a big difference, but for mathematicians it implies that many of the elegant ways to analyze stability break down because these approaches work only with equilibrium points.

Collision of Chaotic Attractors

Basically, the kinds of phenomena that may occur if cycles appear in systems with multiple attractors can also happen if strange attractors meet basin boundaries. However, because of the complex structure of such attractors, the results can be more bizarre. For instance, it can happen in simulations that a multispecies system collapses after a long period of transient chaos, so that in the end, only one or a few species survive.[13,17] In such situations, it may become impossible to predict which of the species will be among the survivors (figure 3.9). The

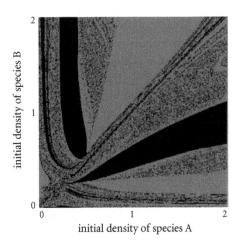

Figure 3.9. Fractal basin boundaries in a multispecies competition model (see text). The three gray tones indicate to which of the different attractors the system will settle if a simulation is started from the corresponding initial condition. Obviously, this outcome is unpredictable for all practical purposes. (From reference 16.)

reason for this complexity is that during the long transient, the system is really moving along the remains of a strange attractor. It is not a true strange attractor, as there are *leaks* through which the system can escape. These leaks correspond to points of basin boundary collision. In this situation, the alternative basins of attraction are intermingled in a fine-grained structure where more and more details are found as one zooms in (*fractal basin boundaries*). This implies that it becomes unpredictable to which attractor the system will settle given an initial state. This is the corollary of sensitivity to initial conditions for a system with multiple attractors.

3.4 Synthesis

We have seen just the tip of the iceberg when it comes to cycles and complex dynamics. It is a fascinating field of research, and especially if one starts playing with models that have complex dynamics on the computer, it is difficult to stop. However, rather than dwelling too much on this vast area of research, I wanted to highlight two aspects that are relevant for understanding critical transitions. First, some cyclic dynamics can be seen as repeated shifts between alternative stable states. Rather than an external change in conditions, such flips are caused by an intrinsic slow change in the system that reverses after the transition, thereby causing a continuous series of flips back and forth. The devastating periodic outbreaks of spruce budworms in boreal forests are a famous example explained further in part II of this book. The second link between the concept of critical transitions and cycles or chaos is that, just like environmental fluctuations and perturbations, such intrinsically generated fluctuations can cause the system to hit the border of an attraction basin, thus invoking a shift to an alternative attractor. Overall, this suggests a richer image of critical transitions. They may be due to changes in external conditions, but intrinsic factors can play a role too. In practice, fluctuations in the state of a system are never of purely intrinsic origin. As discussed in chapter 5, typically a mix of internal mechanisms and external forcing by the weather and other factors drives the fluctuations that may occasionally bring a system over a tipping point for a critical transition.

CHAPTER 4
Emergent Patterns in Complex Systems

The theory presented so far is built on simple models of specific mechanisms that can produce behavior such as cycles and critical transitions. However, similar phenomena may also arise in a way that cannot be understood from the analysis of a few equations or from a simple graphical model. Computer simulations show that striking patterns and dynamics may emerge in networks of units that individually have very simple rules for responding to each other. This emergent behavior cannot be understood from the properties of the units. It can be observed only if we simulate what happens in the entire network. Such emergence is fascinating. In fact, it feels almost like magic if you work with those models. However, it is also in a sense annoying, as it prohibits a straightforward analytical understanding. Nonetheless, it is worth having a closer look at emergent patterns in complex systems, as they seem to correspond to much of what happens in nature. For instance, patterns of movements of ants or schools of fish or flocks of birds cannot be understood simply from looking at the behavior of individuals. Rather, they apparently emerge from interactions in ways that are not so obvious.

Theoreticians working in this field typically search for simple rules for the behavior of the units (for example, individuals, cells, or molecules) that lead to complex phenomena if the units are coupled in a network. Obviously, reality is often different. For instance, human societies or ecological communities are complex systems in which neither the components nor the couplings are simple. Nevertheless, they seem to exhibit many of the hallmarks of *complex systems* predicted by prototype models. Like dynamical systems theory, this is a vast field of research, and I will present merely a bird's-eye view of aspects that are most relevant for the issue of critical transitions.

4.1 Spatial Patterns

Some of the best known results of the work on emergence have to do with the sometimes amazing spatial patterns that can be produced from simple models. I will first highlight a class of systems that was discovered early on and has become the archetype example of emergence. Subsequently, I will briefly explain the phenomena of self-organized criticality and self-organized spatial patterns that appear to be omnipresent in nature once you have developed an eye for them.

CELLULAR AUTOMATA

The idea of emergent complexity can be best illustrated by looking at a class of models known as *cellular automata*. Put simply, cellular automata are like checkerboards in which the squares change color depending on the conditions of their neighbors. Thus, they are regular lattices of cells that interact with their neighbors according to a uniform rule. This rule is the "program" that governs the behavior of the system. All cells apply the rule, over and over, and it is the recursive application of the rule that leads to the remarkable behavior exhibited by many cellular automata. Such a *finite element* approach is also used in more realistic models such as spatial vegetation models described later, as well as for weather prediction, computational fluid dynamics, plasma dynamics, and an almost endless list of other applications. However, the typical cellular automaton research focuses usually on more abstract systems, trying to find out how simple rules on the micro level can generate surprising emergent patterns at the macro level.

A basic example of a cellular automaton is a sheet of graph paper, where each square is a cell, each cell has two possible states (black and white), and the neighbors of a cell are the eight squares touching it. Thus, there are $2^9 = 512$ possible patterns for a cell and its eight neighbors. The rule for the cellular automaton could be given as a table. For each of the 512 possible patterns, the table then states whether the center cell will be black or white on the next time-step. However, usually a simple set of rules is used to determine the next state of the center cell. The "Game of Life" is a famous cellular automaton of this form, and downloading one of the many freeware programs to play with it is certainly the best way to get a feel for what cellular automata are about.

The definition of the game is simple. Each cell can be in one of two states, either "alive" or "dead," and it responds to the state of its eight neighbors according to just four rules:

- Any live cell with fewer than two live neighbors dies of loneliness.
- Any live cell with more than three neighbors dies of crowding.
- Any dead cell with exactly three neighbors comes to life.
- Any live cell with two or three neighbors lives, unchanged, to the next generation.

If you want to program this game yourself rather than download a freeware version to play with, note that all births and deaths occur simultaneously. Together they constitute a single update round of the initial configuration. Thus, the state of the grid evolves in discrete time-steps. The states of all of the cells at one time are taken into account to calculate the states of the cells one time-step later. All of the cells are then updated simultaneously. The basic idea of the "game" is to see if simple initial configurations can be found that generate interesting patterns of life. In a few cases, all living cells vanish, but most initial patterns either reach stable figures, or "still lifes," that cannot change or reach patterns that oscillate forever. There are, however, also rare initial configurations that generate stunning things lake a "gun" (a configuration that repeatedly shoots out moving objects such as the "glider"; see figure 4.1) or a "puffer train" (a configuration that moves but leaves behind a trail of "smoke").

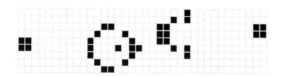

FIGURE 4.1. Initial configuration of the "Game of Life" that evolves into a "gun" that fires gliders eternally.

SELF-ORGANIZED CRITICALITY

In the context of this book, *self-organized criticality* is surely the most important phenomenon that has been discovered in cellular automata. It is well illustrated by the dynamics of the so-called *forest fire model*. Rather than being a realistic description of forest fire mechanisms, models of this type mimic the spatial dynamics of forest fires based on a simple set of rules for the state transitions of grid cells. Many examples can be found on the Internet. The basic idea is that each grid cell can be in one of three states: burning, empty, or occupied by a tree. A visualization typically represents those states as three different colors. At each step:

- Each burning tree becomes an empty site.
- A tree catches fire if at least one of its nearest neighbors is burning.
- An empty site becomes occupied by a tree with probability p.
- A tree without a burning neighbor becomes a burning tree with a small probability f (for example, through a lightning strike).

At the start of this model, trees quickly gain ground. However, after a while, lightning strikes inevitably start fires. The fires spread, destroying large areas of trees. Behind the fires, new trees grow up again. Depending on the chosen parameters, one sees patterns with clusters of trees of different sizes develop and burn. The size distribution of the fires typically follows a so-called *power law*. Small fires are common, and larger fires are increasingly rare in such a way that the logarithm of the fire size is linearly related to the logarithm of the frequency at which fires of that size occur. Interestingly, such power laws are also found for real forest fires (figure 4.2).

The behavior of the forest fire model, in which catastrophes (fires in this case) of different magnitudes interrupt a quiescent state at ir-

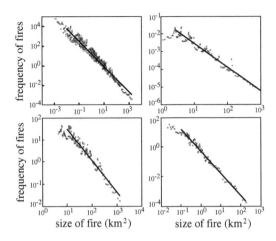

FIGURE 4.2. Frequency-area distributions for actual forest fires and wildfires in the United States and Australia: (a) 4,284 fires on U.S. Fish and Wildlife Service lands (1986–1995); (b) 120 fires in the western United States (1950–1960); (c) 164 fires in Alaskan boreal forests (1990–1991); and (d) 298 fires in Australia (1926–1991) In each case, a reasonably good correlation over many decades is obtained by using the power-law. (From reference 1.)

regular intervals is one of the many examples of what has become known as the phenomenon of self-organized criticality. Essential for such dynamics is that a "tension" or "vulnerability" gradually builds up during the quiescent state that is released through catastrophic transitions. In the case of forest fires, the amount of dead material that serves as fuel represents the growing vulnerability. Numerous systems appear to show a behavior indicative of such a process of self-organized criticality. For instance, earthquakes release tension that builds up in the Earth's crust as a result of tectonic movements, and wars and smaller conflicts release tension in evolving societies. Like forest fires, the size distribution of earthquakes and conflicts indeed tends to follow a power law.

The Danish physicist Per Bak is especially known for developing the theory of self-organized criticality. The wide applicability of the idea is nicely reflected in the ambitious title of Bak's 1997 book *How Nature Works.*[2] The concept was originally developed while studying the theoretical behavior of a sandpile, in which grains of sand are

Figure 4.3. A pile of sand is the epitome of self-organized criticality. Adding grains of sand on the top inevitably leads to occasional avalanches that are mostly small but sometimes big.

sprinkled onto the pile, one at a time (figure 4.3).[3] As the pile grows, its sides become steeper, eventually reaching a critical state when the size of avalanches triggered by added grains follows a power law distribution. The phrase *self-organized criticality* was coined to describe the pile's natural growth to the critical state. The avalanches keep the system in this state as long as sand is added. This self-organizing aspect is important, as it implies that the critical state is reached automatically, without the need to fine-tune parameters. It therefore suggests that the phenomenon may be robust in a wide class of systems.

Simple simulation models can often generate self-organized criticality that mimics natural dynamics beautifully, and it seems reasonable to assume that the basics of many dynamics in nature and society do indeed behave according to this underlying principle. Nonetheless, one should be careful not to jump to the diagnosis of self-organized criticality too fast. In particular, it is important to realize that the fact that the size distribution of events follows a power law does not necessarily imply self-organized criticality. For instance, the size distribution of avalanches of extinctions of organisms in the past as derived from the geological record seems to match a power law, and this has been suggested as evidence for underlying self-organized criticality of the evolutionary process.[4] Indeed, a highly abstract model of evolution may produce self-organized cascades of extinctions of species that obey a power law without an external forcing.[5] However, other mechanisms may generate such patterns as well.[6] The problem is that the size of many elements in nature is distributed according to a power law and that many mechanisms may be responsible for this. Hence, the size distribution of events is not a sufficient diagnostic property.

More convincing cases are built if plausible and realistic models produce apparent self-organized criticality that matches the dynamics observed in the real systems. This is less easy for some systems than for others. In particular, mechanistic quantitative simulation of the dynamics of human societies remains difficult even though the story may seem convincing. For instance, it is well known that small incidents may precipitate large-scale conflicts if the underlying tension is sufficient. The start of World War I, triggered by the assassination of Archduke Francis Ferdinand, which caused an avalanche of conflict with a final dead toll of 10 million when it stopped ten years later, is a case in point. However, as quantitative mechanistic modeling of societies is no easy task, the "explanation" of self-organized criticality remains rather phenomenological. In contrast, self-organized criticality may be shown rather convincingly in a mechanistic model of cliff erosion in tidal marshes, and the results fit observed patterns well for probably the right reasons.[7] In this system, vegetation promotes sedimentation leading to a gradual increase in elevation that increases the vulnerability to cliff-forming erosion due to wave action. Once a cliff is formed, it progresses for some time in an avalanche-like fashion. Clearly, this is in some way quite comparable to the classic sandpile model. All of the processes involved in such systems can be studied experimentally and are rather well understood.

Note that the dynamics addressed by the theory of self-organized criticality are quite close to those addressed by classic catastrophe theory. A slowly changing variable (external or intrinsic) brings the system to a critical point at which a catastrophic transition to a contrasting steady state occurs, usually triggered by a small disturbance (for example, lightning, a wave, or a murder). A major difference is that unlike catastrophe theory that focuses on the interaction of a few variables in a homogeneous system, self-organized criticality focuses on spatially explicit systems in which the same type of catastrophic transition driven by some permanently evolving "tension" occurs over and over again, but always on different places and at different scales. An important aspect of this kind of dynamics is that it is impossible to predict the individual catastrophic transitions. We can perform statistics on their size distributions, but coming back to the sandpile example, one cannot predict whether a particular grain added to the pile will cause an avalanche, and how big that avalanche will be. In a general sense, this is because we are looking at heterogeneous systems in which small events may or may not trigger a chain reaction depending on local tension. Similarly, the size of the resulting avalanche cannot be foreseen, as it depends on the underlying pattern of tension.

Self-Organized Patterns

While most analyses of cellular automata are really fantasy games, self-organized criticality clearly hints at real-world mechanisms. However, there is another class of automata that produces patterns that are much like what is observed in nature. Such models show how self-organized spatial patterns may arise spontaneously in an initially homogeneous environment. Much of this work refers to pattern formation at small scales, such as colored animal skins and chemical reactors. However, recent studies have also stressed the importance of this phenomenon at a landscape level and highlighted the possibility of large-scale critical transitions. Especially, pattern formation in dry landscapes has attracted much attention. In such landscapes, vegetated patches may accumulate nutrients and water from the surrounding barren matrix. This implies a local positive feedback in vegetation development, whereas on a larger spatial scale, vegetation patches compete for resources that can be concentrated.

If one studies the behavior of such spatial models (for formulations, see the appendix, section A.11) over a range of rainfall conditions, fascinating patterns emerge (figure 4.4). Even if the non-spatial vegetation model has no alternative equilibria (panel a), the spatial version may have a range of alternative stable self-organized patterns along the gradient of rainfall levels (figure 4.4b and c). At high rainfall, the landscape is simply covered by a uniform layer of vegetation. Decreasing the rainfall, the uniformly vegetated state becomes unstable (dashed

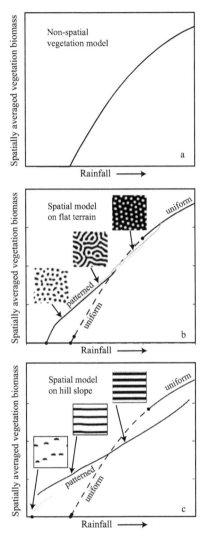

FIGURE 4.4. Implications of spatial self-organization for (modeled) effects of rainfall on the vegetation biomass in dry landscapes. The nonspatial version of the vegetation model (panel a) that is used has no alternative equilibria. In contrast, in the spatially explicit version (panels b and c), the state with a uniform vegetation distribution becomes unstable (dashed line) over a range of rainfall levels, and characteristic self-organized patterns emerge that represent alternative equilibria on a landscape level. The patterns are different for flat (panel b) and hill-slope (panel c) conditions. (Modified from reference 8.)

line in the figure). On a flat landscape (figure 4.4b), a regular pattern of gaps in the vegetation layer arises. Subsequently, a labyrinth pattern arises, and last, a pattern of vegetated dots in a barren matrix is found. All of these states may represent alternative equilibria over some range of rainfall conditions, even though the biomass averaged over the landscape does not differ much between the patterns. At hill slopes where the runoff is directional, the predicted patterns are somewhat different (figure 4.4c). Bands of vegetation perpendicular to the slope are generated. A particularly interesting feature of this and other related models is that at the lowest rainfall levels, a situation with self-enriching vegetation patches is predicted to represent an alternative stable state to a state completely devoid of perennial vegetation.[8–10] The mechanism is intuitively straightforward. The vegetated islands of fertility concentrate water and nutrients locally to a level that allows vegetation persistence even though the average water and nutrient levels in the landscape are too low to support vegetation. Once vegetation patches are lost, the concentration mechanism is gone too. Consequently, recolonization of the barren landscape is possible only if precipitation levels have become high enough to support vegetation in average landscape conditions. Another interesting feature of such models is that they predict that under wetter conditions, a patterned state may be an alternative equilibrium to a uniformly vegetated condition. In this case, the patterned state has a lower biomass (averaged over the landscape) than the uniform state, although on the flat landscape the predicted difference between the two is minor.

Numerous examples of characteristic vegetation patterns that match the self-organized patterns predicted remarkably well have been documented.[8–10] This makes the case for the predicted existence of alternative attractors and hysteresis quite plausible. One of the interesting implications is that the spatial pattern may be used to classify aridity,[8] but also as an indicator of a "near collapse" into the completely barren state. Thus, the pattern may be used as an early warning signal indicating that the vegetated state has a low resilience and that a catastrophic loss of vegetation may be nearby.[10]

Models of self-organized spatial patterns as well as those of self-organized criticality typically address very simple local rules and idealized homogenous initial conditions. In reality, extrinsically determined spatial variation will often be an important confounding aspect. For instance, forests may be more prone to fires in dry spots than in moist valleys, and this may affect the emerging patterns. Also, the assumption that systems consist basically of numerous but identical components that behave according to simple rules is a rather crude simplification in many cases.

4.2 Stability of Complex Interacting Networks

Instead of looking at the dynamics of a spatial grid of cells that interact with their neighbors, one may also study more complex network structures. For instance, large numbers of species interact in nature, and the interaction structure cannot be represented in a simple grid. There is a long tradition of studying models in which large numbers of hypothetical "species" interact. Here I highlight the bit of this work related to alternative attractors and critical transitions.

ALTERNATIVE ATTRACTORS IN COMPLEX COMMUNITIES

It has been shown that large numbers of alternative attractors may arise in complex communities of interacting species. Simulation experiments with computers indicate this,[11] but the phenomenon has also been demonstrated in experiments on microcosms with plankton communities.[12,13] The approach in both modeling and experiments has usually been to add species from a species pool one by one to an experimental community. Some species are able to establish, while others fail upon such experimental invasion attempts. As the number of established species increases, it becomes increasingly unlikely that a new species can enter the community, and eventually a point is reached at which none of the remaining species in the pool can invade the community. At that point, the community is considered a *stable endpoint community*. The interesting aspect from our current perspective is that usually a number of different stable endpoint communities can be formed from the same pool. For instance, suppose that we have a pool of species 1 through 20. It may be that species 3, 5, 9, and 15 form a stable configuration but that also the combination of species 4, 9, 15, and 19 and the combination of species 14, 16, and 20 represent stable endpoints that cannot be invaded. These results are usually discussed in relation to resistance of communities against biological invasions.[14] Also, there is a link to the literature about so-called *assembly rules*.[12] This approach considers communities as a kind of jigsaw puzzle that can be assembled from only certain combinations of species. Some authors have stressed the aspect of "path dependency" or "humpty-dumpy effect," as the presence of different endpoint communities implies that the vagaries of history determine in which community state the system ends up. A related implication is that upon disturbance, a community may settle to a different stable state. Ecotoxicologists who saw that in their experiments have independently coined the phrase *community conditioning hypothesis* to describe the phenomenon.[15]

If one explores the effect of environmental change in models of such communities,[16] the behavior looks much like that of simpler models with alternative attractors discussed earlier in the sense that they may show catastrophic shifts if the environmental conditions pass a certain threshold. (See the appendix, section A.5, for an example of such a model.) Also shifts may be triggered by random fluctuations in the environment, or by the more realistic scenario of a combination of fluctuations around trends (figure 4.5). Good long-term data on community composition are rare, but a detailed reconstruction of changes in the diatom community of a Swiss lake over the last century shows a pattern of occasional sharp transitions that is remarkably well in line with what the models predict (figure 4.6).

Unlike in the models discussed in the preceding sections, these multispecies competition models (like many more complex models) do not allow a simple mapping of the equilibria and their attraction basins. Still, the loss of resilience can be shown through an analysis of the strength of a disturbance needed to invoke a shift to an alternative community state. Since we have many species, the state can only really be plotted in a multidimensional space, if we imagine the abundance of each species to be represented on one axis (for three species, we need a three-dimensional space; for four species, a four-dimensional space; and so on). In such a situation, one needs to be explicit not only about the size of the disturbance, but also about the direction. One

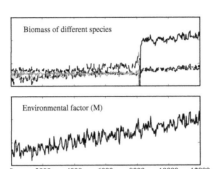

FIGURE 4.5. A model of multispecies competition exposed to fluctuations around a trend in an environmental factor (M) leads to gradual change in communities punctuated by occasional sharp shifts to an alternative community composition. (From reference 16.)

way to do that is to measure disturbance as the displacement in the direction of an alternative attractor. Simulation experiments then illustrate that, indeed, resilience shrinks to nil as one approaches the catastrophic bifurcation (figure 4.7a). Also, simulation experiments in which one gradually increases the environmental factor M until a shift occurs and subsequently slowly reverses the change in M illustrate that, as expected, the model shows hysteresis in response to changing conditions (figure 4.7b).

SELF-ORGANIZED EVOLUTION TOWARD STABLE SPECIES PATTERNS

Although the study of theoretical networks of random species interactions has revealed interesting patterns, the structure of real communities is obviously not random. In fact, models of random species assemblages tend to become unstable if too many species are added, and the problem of stability of complex communities has been fueling debate for half a century already. The question of how species-rich communities can persist may seem a bit academic at times. However, the issue is of obvious importance in view of the rapid loss of species from the planet. Will the hardy species survive and form a resilient community? Should we expect cascades of extinctions if key species are removed?

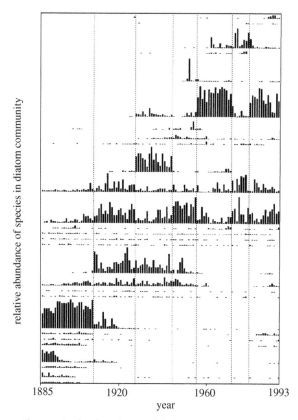

FIGURE 4.6. Changes in the abundance of different diatom species reconstructed from sediment analysis reveal occasional sharp transitions in community composition. (Modified from reference 17.)

Biologists long felt that biodiversity would help to stabilize ecosystems. However, in the 1970s, this intuition was challenged by Robert May, who found that in models, large species numbers caused instability rather than stability.[18] More precisely, he showed that it was difficult to let many species coexist in simple multispecies models. This boosted research revealing why many species can coexist in nature, even if this is difficult to achieve in minimalist models. One of the results was that *connectivity* among species in nature is often relatively small.[19] In a sense, this is another way of saying that species have different *niches*. However, this does not solve the entire problem.

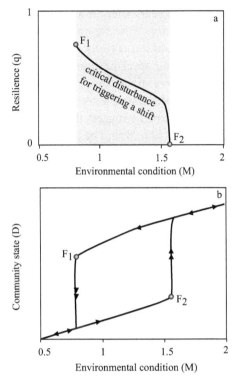

FIGURE 4.7. (a) Over a range of environmental conditions (shaded area), a disturbance beyond a critical size can cause a shift to an alternative community. The disturbance needed (*resilience*) shrinks to nil as change in the environmental condition (*M*) moves the system to a critical bifurcation point (F_2). (b) Gradual increase of the environmental parameter *M* causes the community to shift at a critical condition (F_2); if the change in *M* is subsequently reversed, the system shows hysteresis; and a shift back occurs only when *M* has decreased enough to reach the another bifurcation point (F_1). The community state is represented in a relative way by plotting dissimilarity (D = Euclidean distance) from the initial state (at $M = 5$). (From reference 16.)

In many ways, the root of the quest for explaining biodiversity is a 1959 paper with the intriguing title "Homage to Santa-Rosalia, or why are there so many kinds of animals?"[20] In this classic, the famous ecologist G. Evelyn Hutchinson argues that many species may coexist if they occupy different niches but also points out that the number of species in nature seemed puzzlingly large to be completely explained by niche separation. Two years later, in another classic paper titled "The Paradox of the Plankton,"[21] he drew attention to the fact that the high diversity of phytoplankton is really remarkable, as there seems not much room for niche specialization in this relatively homogeneous environment where all species are competing for a few limiting nutrients and light. Indeed, simple competition models and laboratory competition experiments suggest that the number of species that can coexist in equilibrium cannot be greater than the number of different resources that may become limiting to the growth.[22]

Hutchinson himself had already offered a potential explanation of his paradox suggesting that the plankton community might not be in equilibrium at all: "Twenty years ago in a Naturalists' Symposium, I put forward the idea that the diversity of the phytoplankton was explicable primarily by a permanent failure to achieve equilibrium as the relevant external conditions changed." This nonequilibrium argument remains largely unchallenged, and was confirmed experimentally.[23] Intrinsic chaos can do basically the same job.[24] Essentially, what happens is that by the time a superior species under the given conditions is about to outcompete another one, the conditions have changed enough to make another species superior. This way, many species stay in a game where the rules are, as it where, continuously changed. Another explanation for species coexistence is that ecosystems are not the homogeneous, well-mixed systems represented by most models. Even in seemingly homogeneous environments such as the open ocean, vortices generate transport barriers, preventing complete mixing and therefore promoting coexistence.[25] Last but not least, predators often concentrate on the most abundant prey, and this may help different prey species to coexist, as it prevents any of them from becoming dominant and outcompeting the other.[26] This argument has been expanded recently by showing that natural enemies ranging from specific pathogens and parasites to predators may be responsible for preventing most species from becoming very abundant,[27] thus promoting the possibilities for coexistence.

A quite different angle to explaining the coexistence of similar species is the *neutral theory* championed by Hubbell and others.[28] Their idea sparked considerable controversy.[29] The essential assumption is that all species are basically the same, or at least equivalent, so that no species can outcompete another. Although it may be argued that species sharing an ecological niche and facing the same fundamental trade-offs will coevolve to have roughly the same competitive power,[30] real neutrality is of course an unlikely limit case,[31] and the results have been shown to be quite fragile to relaxation of the assumption.[32]

A solution to the apparent contradiction between neutral theory and niche theory for explaining stable species coexistence is provided by the phenomenon of *self-organized similarity*.[33] This shows that there can be two very different ways for species to coexist: be sufficiently different or be sufficiently similar. With my colleague Egbert van Nes, we hit upon this intriguing finding when we were studying evolution in models in which large numbers of species were competing. To study this problem, we created a hypothetical world in which we first randomly generated species that differed only in one aspect: their position on a *niche axis*. To make this less abstract, we think of this aspect as body size. Each species we generate is assigned a body size. We assumed that species that are more similar in size compete more strongly, as they are more likely to share the same food sources. Initially, we assume that each place on the niche axis (food availability for each body size) is equally good. Species are then created randomly along the size axis and compete with their neighbors depending on how similar in size and abundance they are. Each species is allowed to slowly evolve its body size in the direction where it experiences less competition, and a loss factor that increases with abundance (mimicking a predator focusing on the most abundant species) is added to ameliorate the tendency to competitive exclusion. Although one would intuitively expect that the survivors of this evolutionary game would be species that are equally spread out over the niche axis, the surprising result is that simulations converge to a pattern of self-organized *lumps* that contain multiple coexisting species of similar size (figure 4.8). The distance between such species lumps on the niche axis depends on the niche width of the species in the sense that the lumps are spread farther apart if the standard deviations of the species size distributions (the niches) are broader. Thus, coexistence of different lumps is a straightforward effect of avoidance of competition. However, species that are similar enough escape this rule of limiting similarity and may coexist within the lumps.

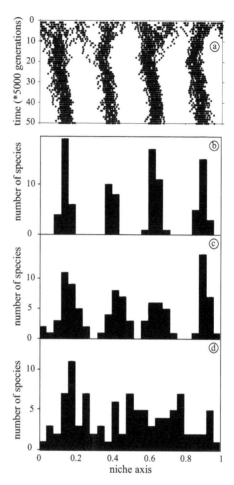

FIGURE 4.8. Simulated evolution of 100 species (dots in panel a) that are initially randomly distributed over the niche axis results in convergence toward self-organized lumps of similar species. Panels (b), (c), and (d) show resulting frequency distributions of species sizes for increasing values of the parameter representing random variation in other factors that affect evolutionary pressure. (From reference 33.)

It seems counterintuitive that coexistence can become easier if two species are more similar. After all, their competition becomes more intense. One way to understand this is to realize that competition becomes close to neutral in the lumps. Species are almost equivalent. Since no species is clearly superior to another, the process of displacement is very slow and can be easily overwhelmed by any of the forces that promote coexistence discussed in the preceding section. So why would we see the curious mix of divergent and convergent competition that creates the lumps (figure 4.8)? The cause of divergence of character to escape competition appears straightforward. However, convergence of species toward other species in a lump that occupies the same niche seems counterintuitive. Nonetheless, it follows from niche theory that as niches of species become more closely packed, positions between species can turn into the worst places in the *fitness landscape*, and evolutionary convergence between close species may occur.[34] Phrased in another way, this theory of self-organized similarity suggests that the lumps of species are alternative evolutionary attractors, whereas the gaps are repellors in niche space. This is well illustrated by the vagaries of the species early in the simulations of evolution (figure 4.8, upper section of panel a).

Empirical support for the mechanism of self-organized similarity comes from striking lumpiness in species size distributions that has been found for many groups of organisms (figure 4.9), ranging from lake plankton and beetles to mammal and bird communities.[35] Clearly, adaptation to preexisting (externally imposed) niches (such as living in the water for whales and fishes) remains an obvious explanation for much of the similarity among species in nature. However, simulations suggest that even if preexisting niches are present, self-organization tends to impose part of the pattern. Thus, it may be a rather robust mechanism. Also, the striking regularity of many of the patterns observed in nature suggests that the mechanism of self-organization may play an important role in many systems.

Surprisingly, self-organized evolution toward attracting lumps of lookalikes may well be common in human societies too. For one thing, size distributions of cities and firms are lumpy too,[36] and clusters of companies, TV channels, and so on seem to provide basically the same services rather than offering something different to consumers.[37] Although different mechanisms may cause lumpy patterns, there is a long line of theory in social sciences starting with the classic work of Hotelling[38] suggesting that lumping may have a similar basis as we found in biological evolution. This line of work shows that competition of entities such as companies or political parties will often lead to convergence rather than differentiation.

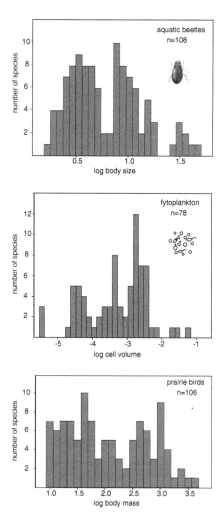

Figure 4.9. Size distributions of species in nature often show a lumpy pattern, illustrated here for European aquatic beetles, phytoplankton species from Dutch lakes, and American prairie birds. (From reference 33.)

4.3 The Adaptive Cycle Theory

There is one other theory that deserves special mention here. It starkly contrasts to all theory presented in this chapter, as it is an inductive rather than a deductive theory. Rather than showing the implications of a particular mechanism, it summarizes a consistent pattern that may be recognized in the dynamics of systems ranging from ecosystems to societies and companies. In fact, it does take a similar niche as the idea of self-organized criticality, as both theories claim to be near-general explanations of jerky change in complex systems. As shown earlier, cellular automata highlight the fascinating aspect of self-organized criticality in an elegant way. However, there is no free lunch, and the price for their elegance is an inevitable superficiality. In the same period that Per Bak worked on self-organized criticality, C. S. (Buzz) Holling felt that there should be rich generic insights to be obtained in the mechanisms of jerky adaptation of complex systems. He inspired a global network of ecologists, economists, and social scientists (*www.resalliance.org*) to work on the broad theme of resilience that he had pioneered earlier,[37] and his efforts of interpreting the numerous case studies converged to a heuristic model that he called the "adaptive cycle."[40,41]

The idea of the adaptive cycle was really rooted in the early work that Holling had done with the mathematician Donald Ludwig and others on the dramatic dynamics of boreal spruce forests that were periodically hit by devastating budworm outbreaks.[42] However, over the years, numerous examples of systems that went through occasional phases of destruction and reorganization convinced him that those were really essential aspects that were too often neglected in ecological theory. The adaptive cycle is a heuristic model that puts emphasis on these aspects. Rather than being precise and mathematical, it has a holistic character. Although it has been conceived by an ecologist, it has proved an inspiring tool for thought, especially for social scientists. Perhaps its attractiveness to social scientists is in part because they are used to the inductive approach underlying the adaptive cycle model. In any case, this broad enthusiasm suggests that the model makes sense of important aspects of social dynamics in a novel way and fills a niche that was left empty by rigorous mathematical ap-

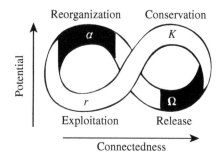

FIGURE 4.10. The *adaptive cycle* model of change emphasizing role of episodes of destruction and reorganization that are characteristic of the dynamics of many living systems. (From reference 41.)

proaches proposed earlier. It is therefore an interesting complement to the models we have seen so far in this book. An accessible account of the ideas can be found in the book *Resilience Thinking*.[43]

The idea is that living systems tend to cycle through four principal phases: exploitation (r), consolidation (K), destruction (ω), and reorganization (α) (figure 4.10). For an ecological example, think of the long-term dynamics of a forest. Slow succession allows it to develop from pioneer species (r) to climax phase with large trees and well-adapted undergrowth species (K) until a major disturbance such as fire or pest (ω) destroys it. As in the case studies of self-organized criticality, one can imagine that such destruction becomes increasingly likely as biomass and dead organic matter ("fuel" for pests or fires) accumulates and the system becomes gradually more vulnerable. Eventually, the disturbance releases accumulated nutrients, allows light to reach the bottom, and triggers a phase of reorganization into the start of a new cycle (α). The reorganization phase may be seen as a window of opportunity for novel developments, as new species may get a chance that they were unable to enter during the climax phase.

Thus, a pattern of two phases of growth is generated, followed by two phases of reorganization. The first two form a familiar, slow, fairly predictable pattern of growth called the *forward loop*; the second two constitute the less familiar aspect emphasized by this model, an unpredictable and often rapid *back loop* of reorganization. It is the two together that make the cycle adaptive. Novel elements can accumulate, largely unexpressed, during the forward loop. Then, in the back loop, they become the seeds for the novel combinations that launch the next cycle. Holling emphasizes that adaptive cycles occur at all scales, in ecosystems ranging from leaves to biomes, from centimeters and days to hundreds of kilometers and millennia. As in self-organized criticality, transformations at small scales may trigger cascading avalanches to larger scales. At the same time, local dynamics are usually prevented from moving in all too novel directions by the memory of large-scale properties such as regulation of the regional climate by forest, seed stores, and mobile species.

Such adaptive cycles can be recognized not only in the dynamics of ecosystems. Once you have developed an eye for them, you see them everywhere. The book *Panarchy*[41] interprets numerous examples in this fashion, much as the book *How Nature Works*[2] does for self-organized criticality. Paradigm shifts in science, dynamics of businesses in developing markets, and crises in societies all may be mapped onto the adaptive cycle model with little effort. An important contribution of the adaptive cycle model is its emphasis on the fact that crises open a window of opportunity for novelty. Such novelty is suppressed during the more rigid phases when growth has resulted in established structures that leave little room for change. As discussed later in more detail, the dynamics of learning and innovation are much more important in social systems than in ecosystems, where similar processes happen essentially on an evolutionary timescale. The decline of dinosaurs opened a window of opportunity for the radiation of mammals, but that process took millions of years. In human societies, phases of apparent equilibrium are punctuated by swift transformations, of which examples are numerous.

In episodes of crisis, there is space for reorganization and innovation. Although that sounds attractive, there is a flip side to the coin. Uncertainty is high, control is weak and confused, and problems are often too complex to see clearly how they can be solved. As discussed in section 12.4, there is a niche for new "sense-making" in such complex situations of crisis, and a charismatic leader may cause the precipitation of confused feelings into a comprehensive new worldview. It is therefore a time when individuals have the greatest chance to influence events, and a time when not only a Gandhi but also a Hitler can shape the future. More generally, the often-heard opinion that "timing is crucial" in good management and politics may be seen as referring to the window of opportunity offered by the reorganization phase in the Hollings model.

Taking another perspective, the apparently crucial role of the *back loop* in many systems makes one wonder whether such dynamics might be not only used but also intentionally induced. Indeed, it seems not uncommon that effective leaders first induced a crisis, and subsequently used it. Also, it is common practice in some fields to create small crises in order to prevent big ones. For instance, forest managers regularly induce small controlled fires. For a long time, the dominant management policy was instead to extinguish all fires as fast as possible in order to prevent damage to the forest. However, under such a regime, an increasing amount of dead organic material is accumulated in the forest to the extent that any spark may trigger a wildfire of devastating scale and intensity. Intentionally triggered small fires prevent such large catastrophic events. Analogously, it may be argued that it may be good to promote small cycles of adaptive innovation in social systems to prevent massive transformations that can have negative and unpredictable effects. Indeed, a common management philosophy is that it is good to "shake the tree" every once in a while, and most companies actively promote a process of innovation.

The problem is that there is a fundamental trade-off between being adaptive and being efficient. It is simply not easy to have the best of both worlds: the self-organized optimization of performance of the system in a *K*-phase of the Hollings cycle goes inevitably with rigidity. One might see the *forward loop* in the Hollings cycle as a process of digging a deep basin of attraction. The fact that the basin is deep corresponds to a strong performance. However, it also keeps the system as it is in place in a rigid way. Slow larger-scale developments inevitably change the overall stability landscape in such a way that this basin of attraction ends up being a tiny valley in some scary high place. Eventually, a large catastrophic transformation out of this situation is inevitable. If the self-dug stability basin had been less deep, changes in the stability landscape might have led to earlier adaptation and resettlement into another place. Large companies that are successful in the long run deal with this trade-off by having research and development departments that focus not only on optimization of the current practice (digging the valley deeper) but also on long-shot explorative research that aims at probing for remote good places in the fitness landscape to which the company can shift before the current position will lead to a catastrophic collapse. I discuss the issue of adaptive capacity and rigidity in societies in deeper detail in section 12.6.

Depending on your scientific background, you may feel that at this point, the discussion has become rather loose and informal. These are intuitive metaphors rather than rigorous models. Indeed, in contrast to the models discussed in this book up to now, the adaptive cycle is a heuristic model, obtained in an inductive way. It is an attempt to capture the essence of patterns observed in many different case studies. Such an approach is common in the social sciences, but not in the natural sciences. Interestingly, ecology is a branch of science where both approaches are used to a certain extent. Still, it is far easier to publish an article about an uninteresting but heavily replicated experiment in glass jars than about the effects of a large-scale experimental manipulation of a lake that tests an interesting hypothesis but has only a single imperfect control. Such an emphasis on controlled experiments and reductionism may be good for rigor, but it can also hamper our exploration of mechanisms that drive large-scale dynamics.[44] I believe that to unravel the mechanisms that govern the dynamics of large complex systems, it pays off to every now and then explore holistic views suggested by large-scale experiments and inductive analysis of patterns. These can point to phenomena and mechanisms that we would never have imagined following the *local search algorithms* of deductive science. Once identified, such potentially interesting possibilities can be further scrutinized using controlled experiments and models.

4.4 Synthesis

The work highlighted in this chapter shows that critical transitions may not always be understood from the simple kind of models capturing the interaction between two or three variables shown in the previous chapters. The holistic adaptive cycle theory is one example suggesting a pattern of transitions that is obvious in many systems but is hard to capture in a few equations. On the reductionistic side, much work has been done showing how *complexity* may arise in systems of many simple components interacting in simple ways. Cellular automata have revealed how systems under a continuous pressure may eventually self-organize into a permanent critical state where the pressure is released through critical transitions that are individually unpredictable, but that collectively obey a probability distribution of sizes that follows a power law. This behavior is an emergent property of the system as a whole and cannot be understood by looking just at the rules of the game of how the cells of the automaton interact. Other models of numerous interacting components suggest how fascinating spatial patterns may self-organize in live systems ranging from animal skins to desert vegetation, and how species may self-organize in groups of lookalikes over evolutionary timescales. Such patterns are linked to critical transitions in two ways. First, the pattern may represent a series of self-organized alternative attractors. For instance, species in the simulated evolution move toward the attracting lumps of similar species and are repelled by the gaps. Second, as external conditions change, the self-organized pattern may go through a critical transition to an alternative stable self-organized pattern or complete collapse. For instance, as rainfall decreases beyond a critical level, the self-organized pattern of desert vegetation is predicted to vanish, and recolonization of the resulting completely barren state requires much more rain than was needed to maintain the patterned state.

CHAPTER 5

Implications of Fluctuations, Heterogeneity, and Diversity

The brief excursion through theories in the previous chapters may leave you with the impression that all phenomena that can occur in complex systems are in principle understood and classified. However, with the exception of the holistic adaptive cycle theory, most of the work up to now deals only with the easy part: minimal models simulating the behavior of caricatures of complex systems in a homogeneous and constant environment. Reality is something quite more complicated. Does this mean that the models are of little use, or would their predictions be robust in the sense that they would still hold in a messier world than the idealized one in which they were analyzed? As you will see now, real-world complexity may in some cases smother and buffer the characteristic phenomena illustrated in the previous chapters, whereas in other situations, stability shifts, cycles, intrinsic chaos, and self-organized patterns may still stand out clearly in the overall dynamics observed in reality. We will look into some essential aspects that make reality often different from what is found in the simple models: *permanently changing conditions, spatial heterogeneity*, and the *large number and diversity of components* in dynamic interactions. Admittedly, science is less advanced here. While in many cases, we can see the contours of how these complications should affect the big picture, in other cases, we hit the limits of where theoreticians have settled, offering views of vast unexplored fields of research.

5.1 Permanent Change

One of the obvious simplifications behind most models of dynamical systems is the assumption that conditions remain essentially constant (or at the most periodically oscillating). This allows us to think of attractors and bifurcations and makes it possible to systematically explore all the possible *asymptotic modes of behavior* to which the model can converge. Alas, most situations in the real world are quite different from that ideal picture. Conditions are always changing. We may arbitrarily distinguish two aspects of such change: slow change in baseline conditions and stochastic environmental fluctuations driven directly or indirectly by the weather and other mechanisms.

OMNIPRESENT SLOW CHANGE

Ecosystems or societies are obviously never stable in the sense that they do not change. There are always slow trends. For instance, technology evolves, species evolve, lakes fill up with sediment to eventually become land, and over long timescales, natural climatic change has gradually turned moist places into deserts and tropical seas into tundra. Basically, the question of whether one wants to address such shifting baseline conditions depends on the timescales of interest. For instance, the population cycles between zooplankton and their algal food take weeks. Obviously, for unraveling the mechanisms that drive these dynamics, it makes little sense to worry about the fact that the lake will fill up with sediment over the coming millennia. Also, the slow variations in the Earth's orbit such as the 40,000-year cycle in the tilt can easily be neglected for most process studies. On the other hand, the relatively abrupt desertification of the Sahara region about 5,500 years ago has probably been driven by such subtle slow change in the Earth's orbit. The bottom line is of course that if one wants to unravel the mechanisms that drive certain dynamics in a system, the challenge is to leave out the things that do not matter too much and focus on the major factors. That is what modeling is about, and that is what theory is about: constructing simplified images of the world that help to see the forest through the trees in this dazzlingly complex world.

A direct application of this philosophy with respect to timescales in modeling is the so-called slow–fast separation (section 3.1). The essence is that for studying the dynamics, one focuses only on a particular timescale. Factors that change very slowly relative to that timescale can be considered constant for simplicity. On the other hand, processes that are very fast relative to the scale of interest can be considered to be in an instantaneously adjusting equilibrium called *quasi steady state.*

Dynamic Regimes Rather Than Clean Attractors

Less easy to deal with are fluctuations in environmental factors that happen on the same timescale as the dynamics one wants to study. For instance, natural populations always fluctuate, and this is probably largely due in part to seasonality and fluctuations in weather. On the other hand, even if environmental conditions were constant, populations would probably remain fluctuating. This is deduced mainly from the fact that models of interacting populations often converge to such cycles or chaotic dynamics rather than to stable states (see chapter 3). Also, some experimental work on plankton in microcosms shows that even if conditions are kept constant, a system of interacting species may keep fluctuating wildly.[1] If one looks at natural time-series, the role of such intrinsically generated dynamics and the effect of external forcing remain difficult to unravel in spite of ever-increasing sophistication in statistical techniques.[2]

Whatever the relative importance of intrinsic and extrinsic dynamics is, fluctuations rather than stable states are obviously the rule in nature. It is important to choose terminology well in this respect. Clearly "stable states" do not exist in reality. Therefore, if one refers to real systems rather than models, it is wiser to use words such as *regimes* or *attractors* instead of the terms *stable states* or *equilibria,* which seem to exclude dynamics. In fact, *attractor* is still a technical modeling term that includes cycles, quasi-periodic behavior, and chaotic attractors but does not assume an effect of randomly fluctuating conditions. This leaves *regime* as the most realistic term for what the state of most real systems in nature and society is. Sudden changes between regimes are then *regime shifts*.[3] While this term is indeed widely used now, it may refer to any sudden change, including those caused simply by a sudden change to a different set of external conditions. Critical transitions are the subset where a system is pushed over a threshold where a positive feedback causes a self-propagating shift to an alternative regime.

Long Transients

Clearly, things become much messier now than they were in the clean theoretical world sketched in the previous chapters. Most theoreticians stay away from the murky waters of stochastically forced systems, and for "theoretically good" reasons. Analytical, clean, and general results are much less easy to obtain in noisy models. So how important are all these deviations from the theoretical framework of attractors and bifurcations really? That depends. In many situations, a simplified noise-free image helps a lot to see the big picture. On the other hand, many systems may be in transient states and possibly far from their theoretical attractors most of the time.[4] In models, it can be seen that the time that systems stay away from attractors can be boosted if there are places where they tend to "hang around," even if they do not represent attractors. Examples of such hang-around places are saddle points, remains of strange attractors, or ghosts of equilibria that disappeared after a fold bifurcation. Because model systems may linger for long periods around unstable points or cycles, one would suspect that this kind of behavior could well dominate dynamics in nature too. So far, few studies have looked for such prolonged transient dynamics. One example is a laboratory experiment with flour beetles showing that the system may stay close to an unstable point for some time before spiraling out to a stable limit cycle.[5] Another example is the tendency of shallow lakes to stick to an unstable clear state for years before sliding into the stable turbid state.[6] The idea here is that occasional droughts result in near desiccation of the lakes. This kills most of the fish and pushes the system into a clear situation that is not stable, but is almost so. This situation thus represents a *ghost* of a stable state, and the dynamics away from it are very slow.

5.2 Spatial Heterogeneity and Modularity

In addition to the narrow focus on attractors and the neglect of permanent change, an important oversimplification in many of the classic models that are used to study stability is the neglect of spatial heterogeneity. Looking at a landscape, one usually sees places that seem more fertile, wetter, or more protected than other places. The difference between sites may be gradual, but sometimes it is sharp enough to lead to a patchwork of suitable and unsuitable places for many species. In fact, many forests, wetlands, coral reefs, and other ecosystems consist of patches of habitats. Such patches are connected by exchange of matter and organisms. Some of this exchange is passive, but many organisms also actively move between areas. Intuitively, it seems obvious that such spatial heterogeneity must in some way affect the likelihood that we see phenomena like cycles or sharp shifts. One would expect things to be somehow buffered in spatial heterogeneous settings. Indeed, several authors have stressed the possible importance of spatial heterogeneity and exchange for alternative stable states.[7] However, few studies have really explored the topic experimentally or with models.

EFFECTS OF SPATIAL STRUCTURE ON OSCILLATIONS

Although many predator–prey models show wild cycles, natural populations rarely do. One of the most frequently mentioned topics in the literature about stabilization of prey–predator dynamics is spatial heterogeneity. Using models, many authors have shown a stabilizing effect of partial isolation of habitat patches.[8] Other models have been formulated to show that prey–predator oscillations are reduced or eliminated when the predators aggregate in patches with high prey density.[9] Stabilization can also be achieved by limiting the speed of movement of individuals in individual-based prey–predator models.[10] All of these mechanisms are in fact closely related. The space outside the patches where the predator is concentrated can be considered a refuge where part of the prey population can escape predation. In the appendix (section A.8), I give a simple model where the effect can be seen for the case of lake plankton. At first glance, a counterintuitive result from this type of study is that whether an oscillating system can be stabilized by spatial structure depends critically on the connectivity of the "compartments." In complete isolation, each of the spatial sections of the system simply behaves in its own way (for example, cycling). On the other hand, if spatial exchange is very strong, spatial structure becomes irrelevant, as the entire system behaves in synchrony (cycling again). Therefore, the strongest stabilizing effect is found at intermediate levels of spatial coupling (no cycles).

Implications of Environmental Heterogeneity for Stability Shifts

The example in the previous section illustrates that it is not only the heterogeneity of the environment itself that matters, but also the extent to which the separate parts are coupled. Obviously, this is also true for the chance that we see catastrophic shifts in the system. For instance, imagine a dry landscape in which vegetation may be present or absent depending on the general climatic conditions but also on how good the soil and other local conditions are on a specific *microsite*. (This problem is also discussed in section 11.10.) If water is already a major limiting factor and the climate becomes drier, vegetation may decline. However, in the heterogeneous landscape, some sites are better than others. Consequently, vegetation will hang on longer in, say, relatively moist and fertile valleys than on sun-exposed dry hills. This implies that even if on a given spot vegetation would collapse completely if precipitation falls below some critical threshold, the total amount of vegetation seen on a larger scale in the landscape will decline gradually if the climate becomes drier, as progressively more sites become unsuitable for plant growth. Thus, even if on a small scale, there are sharp collapses, on a larger scale, these average out to result in a more gradual response. This is fairly straightforward so far. However, simple reasoning does not work anymore if there is some coupling between the sites. This is the usual situation in practice.

For example, vegetation in dry areas may promote rainfall in the region. This implies that plants on all sites benefit from the presence of vegetation on other sites, ameliorating the regional weather conditions. This may lead to alternative stable states and sharp shifts on a landscape scale. The birth of the Sahara desert and a potential shift to a drier state in the Amazon are examples of the possible effects of such a feedback that couples local sites on a larger scale (see section 11.1).

Perhaps more common are situations in which the coupling has a local character. For instance, dispersal of seeds from nearby patches may enhance the likelihood that an empty patch becomes vegetated. Similarly, in shallow lakes, submerged plants enhance clarity of the surrounding water, which is good for light conditions and plant growth. This can be a very local phenomenon. Often water is crystal clear in submerged plant beds but very turbid in nearby unvegetated parts of the lake.[11] Still, depending on the water movement, there is obviously some spatial coupling. Vegetated sites will suffer a bit from the inflow of turbid water. Similarly, unvegetated sites will receive some clear water from the plant beds, which may enhance the possibility that they shift to a vegetated state.

The big question is how such local effects play out on a larger scale. In some situations, one could imagine a domino effect. If one site flips to another state, it triggers its neighbors to flip, and so on, causing the entire landscape to shift to an alternative state. However, sticking with the same analogy, in other situations, the dominoes may be out of reach of falling neighbors or may simply stand too firmly to be toppled over.

We can explore the potential implications of spatial exchange and heterogeneity by placing a model with alternative stable states in a spatial context: define a grid of patches, and then run the model in each grid cell, allowing some exchange between the cells and ensuring that local conditions differ between the grid cells to mimic the heterogeneity of the landscape. Several simple models have been analyzed in such a way.[12]

As an example of what comes out of such analyses, consider the results of a spatial model of submerged vegetation in a lake (see the appendix, section A.12, for equations). The simplest version of the model considers a single row of grid cells. Each grid cell has a different depth. This affects the suitability for plant growth, because it is increasingly more difficult for the rooted plants to grow if the water becomes deeper and less light reaches their leaves. We study two types of situations: a lake with a smooth depth gradient and a lake with a more bizarre depth profile in which neighboring cells have randomly varying depths. A parameter (d) mimicking the dispersive movement of water regulates the coupling between neighboring cells. We now study what happens if the lake is exposed to change in the nutrient level (figure 5.1). This causes an increase in phytoplankton growth, making the water more turbid and life more difficult for the submerged plants (section 7.1). The nutrient level in the model is set by a parameter that represents the turbidity in the absence of vegetation.

The mother-model represents a *homogeneous lake* in which all sites have the same water depth. As all sites are equal, this situation can be studied by looking at just a single grid cell, and the behavior simply corresponds to the minimal model in which no spatial aspects are considered (figure 5.1a). This lake shows a classic hysteresis pattern, shifting between a vegetated and an unvegetated state at different critical nutrient levels. As a next step, consider a lake in which the different grid cells have a different depth but operate as completely segregated, as there is no dispersion of water in the lake (figures 5.1b and 5.1c). Now the response of the lake to change in the nutrient level becomes more gradual. This is because the deepest spots lose their vegetation earlier than the shallower spots where light can still reach the plants even if the water becomes more turbid. Note that hysteresis remains. In each of the isolated grid cells, the critical nutrient levels for a *forward* or a *backward* switch differ. Therefore, if the nutrient level in the lake is reduced, the path to recovery of the vegetation differs from the trajectory of vegetation loss.

Now consider a well-mixed lake in which the water is exchanged vigorously between neighboring cells. The response pattern then depends strongly on the spatial correlation of the environmental condition of the sites. In the bizarre lake where depth varies randomly between cells (figure 5.1e), we are back to pronounced hysteresis and sharp synchronized catastrophic shifts in response to changing nutrient levels. In contrast, if there is a smooth depth gradient (figure 5.1d), the response of the lake remains more gradual, and hysteresis becomes reduced. In this case, a domino effect is important. In a lake that is completely vegetated or completely devoid of vegetation, it takes a long time for the first cell to shift to a deviating state. However, as soon as it flips to the other equilibrium, it pushes the neighboring cells over too.

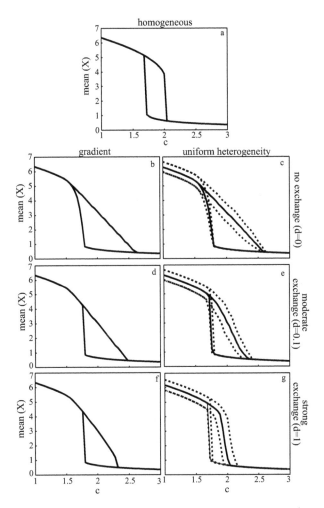

FIGURE 5.1. The effect of spatial heterogeneity on hysteresis in a simple model in response to a parameter (c). (a) Hysteresis in the original model without environmental heterogeneity; (b) average response of the system if the model is run over a smooth environmental gradient of cells without dispersion; (c) the same, but now in an environment that varies randomly from place to place (0.05 [dotted line], 0.5 [solid line], and 0.95 [dotted line] percentiles of 100 runs); (d) gradient with moderate dispersion; (e) random heterogeneity with moderate dispersion; (f) gradient with strong dispersion; and (g) random heterogeneity with strong dispersion. In the randomly varying landscapes (figures on the right), there is also an indication of the distribution of the results looking over 100 runs: 0.05 (dotted line), 0.5 (solid line), and 0.95 (dotted line) percentiles. (From reference 12.)

Although the lake example may seem rather specific, analysis of other models suggests that the main aspects of the behavior of this model appear quite general.[12]

In summary, in landscapes that are spatially heterogeneous, a local tendency to alternative stable states tends to smooth out on a larger scale, but hysteresis can be preserved on the larger scale in the sense that the system follows a different path in its response to an increase and decrease of a control factor. Also, if spatial coupling is strong enough, even heterogeneous landscapes can show catastrophic shifts synchronized over large scales, especially if the landscape heterogeneity does not take the form of a smooth gradient.

5.3 Diversity of Players

Although some systems consist of numerous similar elements, allowing approaches to search for emergent behavior as discussed in section 4, the components of most complex systems such as societies and ecosystems are rather diverse. The classic dynamical systems approach to study stability in such systems concentrating on only a small subset of variables is problematic too. It is true that in some ecosystems, a few species (or functional groups) are really the "drivers" that govern the dynamics, whereas the rest are just "passengers" that fill the gaps and follow the movements but do not determine the direction. In such situations, a focus on the driver-species is obviously useful for unraveling the mechanisms behind the behavior. Although we have learned a lot from such simplifications, the question always remains what we lose by leaving out the large variety of components in complex systems. The discussion has been particularly strong around the issue of biodiversity, as the rapid loss of species that we face today has led to the question of whether ecosystems really need all these species to function properly.[13]

With respect to biodiversity research, the focus of interest has largely turned back to a question that really was at the root of the early debate: does biodiversity enhance stability? A closer look at the theoretical and empirical studies of diversity–stability relationships reveals that this is not a well-posed question.[14] The problem is that biodiversity as well as stability can be defined in many ways, and the answer to the question of whether diversity promotes stability can be positive or negative depending on the definitions. Therefore, one should focus on more precise subproblems. In view of the focus of this book, it is especially interesting to explore what the implications of biodiversity are for the probability of large stability shifts.

Today, two main ideas on the importance of biodiversity for ecosystem functioning dominate the literature. First, and very closely related to the early ideas,[15] there is the *insurance* hypothesis,[16] which states that the functioning of diverse communities will remain more stable in the face of environmental fluctuations and disturbances. Model analyses, but also intuitively straightforward reasoning, tell us that if there are more species that can perform a certain functional role in the ecosystem, the consequences of one species running into trouble for the system will be less (figure 5.2). Second, it has been argued that communities with more species tend to perform better in aspects such as biomass production and nutrient recycling. This is due in part to the *complementarity* between species, making the group perform a certain task better than a monoculture of any of the species, and in part to the statistical effect that with more species present, there is simply a higher chance of having one that performs particularly well.[17]

With respect to the question of how loss of biodiversity could affect the risk of ecosystem regime shifts, it is useful to distinguish between diversity in the way species respond to stressors (for example, sensitivity to cold spells) and diversity in their functional roles in the system (for example, which plants they eat). Complementarity refers to *functional diversity*, whereas insurance has to do with *response diversity*.[18]

Let us first look at the possible implications of response diversity for stability. Suppose that we have a key functional group for maintaining the current ecosystem state, such as herbivores that control macro-algae on coral reefs. Now suppose that the species in this functional group do not differ in their functional capacities but do differ in their sensitivity to stressors. Thus, we have high response diversity, but no functional diversity. Clearly, loss of species in this group would reduce the probability of the group as a whole sustaining its functional role of algal control. There is simply less effective insurance

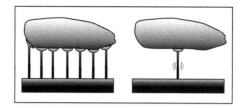

FIGURE 5.2. Intuitive representation of how the loss of species can make an ecosystem fragile. If only one species remains to represent a key functional group, loss of that species through a disease outbreak or other problem can induce an ecological collapse.

against adverse impacts. However, as the functional capacity of all species is equal, loss of species would not affect the way in which algal biomass responds to an increase in nutrients. All herbivores can perform the control equally well. One way to characterize such a situation would be to say that the herbivore-controlled state of the reef has a higher *resistance*, as the key functional group will be less affected by a given random disturbance. Graphically, this can be depicted as an increased depth of the attraction basin (figure 5.3). Indeed, the pathogen that wiped out sea urchins would probably not have led to regime shift of the Caribbean reefs into an algal-overgrown state if herbivorous fishes had not been eliminated from the herbivore guild by overfishing.[19] Note that it is response diversity, not species diversity that matters for building resistance. For instance, in the 1890s, rinderpest killed 80% to 90% of the domestic herbivores as well as wild ungulates in Africa, leading to a large-scale shift to a forested state.[20] The fact that there were many species of herbivores did not help, as they were all sensitive to the pathogen.

Now consider the role of functional diversity. In practice, species will often differ in their functional capacity. There may be some complementarity in function, and some species may simply be more powerful agents with respect to certain functions than other species. If species in a functional group are collectively involved in a critical feedback for maintaining the current ecosystem state, loss of part of the species may reduce resilience. For instance, in certain dry regions of the world such as parts of Amazonia, vegetation cover may promote precipitation. Because of the dependence of vegetation on rain-

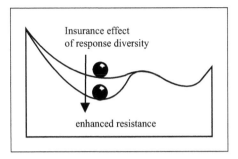

FIGURE 5.3. As diversity within a key functional group for maintaining a particular stable state is larger, it becomes more likely that some of those species happen to be able to maintain the functional role in the face of certain adverse impacts. This insurance effect of biodiversity might be loosely symbolized by increasing the depth of the basin of attraction, representing the fact that more disturbing force is needed to kick the system into the alternative basin.

fall, loss of vegetation in such regions may lead to a self-propagating collapse into a dry state.[21] Not all plants are equally effective in maintaining the feedback on the water cycle. Deep-rooting trees can perform an especially critical role, as they recycle deeper groundwater into the atmosphere. Thus, loss of species in this key plant group with deforestation increases the risk that incidental dry years invoke a collapse into a regional drier state.[22]

Obviously, biodiversity loss is not a random process. Species that are most sensitive to certain stress factors will disappear first. Therefore, the correlation between response traits (that is, sensitivity) and functional traits should affect the way in which ecosystem resistance and resilience changes with species loss.[23] Much remains to be explored in this field. Intuitively, one would expect that loss of the most-sensitive species results in a relatively hardy remaining community. However, this is not necessarily the case in communities of interactive species.[24] Models suggest that such communities are characterized by nonlinear unpredictable response to environmental change.[25]

Importantly, loss of species almost inevitably leads to an increase in abundance of the remaining species in a functional group. For instance, overexploitation of fish on Caribbean reefs resulted in extremely high abundances of sea urchins,[19] and overexploitation of various fish stocks in the gulf of Maine may well have allowed lobster and crab populations (alternative top predators) to skyrocket.[26] In a wider sense, crop systems and monospecific forest stands are also examples. The resulting high population density of a single dominant remaining species is well known to increase the risk of an outbreak of disease or pest.[27]

In fact, disease outbreaks can themselves be considered as regime shifts that happen when a certain critical threshold is passed. However, while the direct effects of the epidemic can be short-lived, an epidemic in a key species can imply a disturbance that pushes the entire ecosystem over a threshold to an alternative state from which recovery is very difficult. The disease that caused mass mortality of sea urchins in Caribbean reefs is an obvious case in point (see section 10.20). Loss of the sea urchins resulted in massive overgrowth of the reefs by macro-algae that remained in this dominant position for decades. Extreme abundances of species in systems with few species may be further boosted by increased nutrient load to the system. Thus, species loss and eutrophication may act in synergy to increase the risk of epidemics.

In summary, there are several reasons to expect that loss of species increases the risk of sudden regime shifts in ecosystems. Theoretically, this follows from the combination of current theories on catastrophic ecosystem shifts and theories on biodiversity, and the case of coral reef collapse illustrates the point in a dramatic way. Although well-documented case studies that link regime shifts to a biodiversity-related loss of resilience and resistance are still rare, there is in fact general empirical evidence that strongly supports this view. Conspicuous population collapses and cycles are long known to be more common in systems that naturally have fewer species, such as islands and high-latitude ecosystems.[28] Kelp forests are a good case in point (see section 10.2). Kelp forests along the Northeast coasts of the United States are subject to occasional massive long-lasting collapse of kelp cover. In contrast, the comparable but much more species-rich kelp forests along the Californian coast quickly recover on disturbances.[29] An obvious corollary is that sudden dramatic shifts in ecosystems may well become more common in a world that has fewer species.

In spite of fundamental differences between human societies and ecosystems, some parallels in the relationship between stability and diversity seem obvious. Surely, the insurance principle can be found on all levels in societies. Families may spread the risk of losing their livelihood under unpredictable conditions by having a diverse portfolio of ways to make a living, companies with a diverse portfolio are less vulnerable to market vagaries, and societies with a more diverse pattern of resource use, activities, and lifestyles may be more resilient to disasters.[30] It might therefore be argued that governance of disaster-prone societies should be aimed at creating such diversity. Also, adaptive capacity and innovation are clearly good things for societies facing major change. A simple model of critical transitions in attitude suggests that diversity among individuals reduces the risk of groups or societies to become locked into inertia.[31] However, the mechanisms that drive diversity, adaptive capacity, and innovation in social systems are obviously more complex than what can be captured in such a simple model. I come back to this theme in chapter 12.

5.4 Synthesis

Clearly, the image of critical transitions becomes more complicated if we move from simple models to consider some of the key ingredients of the messiness of the real world, such as environmental turmoil, spatial heterogeneity, and diversity. In view of the omnipresent turmoil and change, surely the concept of *stable states* is too simple to capture almost any real system. However, as long as fluctuations are not too wild and one focuses on the right timescale, this does not preclude the possibility that critical transitions occasionally overwhelm the system. Relatively little systematic work has been done when it comes to the implications of diversity, and there appear to be many subtleties. However, the emerging pattern seems to be that heterogeneous systems that consist of relatively isolated units and have a large diversity of players are less likely to go through occasional large-scale sharp transitions. This is especially relevant if we look at it from the other side: loss of diversity and increase of connectivity may increase the tendency to see occasional critical transitions in complex systems.

CHAPTER 6

Conclusion: From Theoretical Concepts to Reality

In view of the big gap between most real-world situations and the simple models on which most theory is based, it is perhaps not surprising that much confusion has arisen about the interpretation of some concepts. There is simply an unavoidable tension between the clear but sterile and rigid definitions from dynamical systems theory and the rich spectrum of behavior observed in real complex systems. The following sections highlight the nuances needed to make practical use of five key concepts: *alternative stable states, stability basins, resilience, adaptive capacity,* and *critical transitions.*

6.1 Alternative Stable States

There are two major sources of confusion about the concept of alternative stable states. First, some debate has been caused by the sterile concept of a *stable state.* In models, there can also be alternative cycles or strange attractors, and in the real world, environmental fluctuations and trends mix with intrinsic dynamics to create everlasting turmoil. One way to cover all this is to use the term *dynamic regime* instead of *stable state.*

In addition to the problematic interpretation of the concept of stable state, there is a second, much trickier source of confusion. In

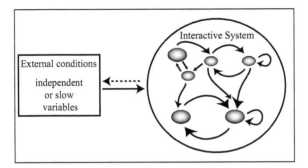

Figure 6.1. The concept of alternative stable states refers to the situation in which a system can be in more than one contrasting state under the same external conditions. Those external conditions consist of factors that are not notably affected by the system or that change only very slowly compared to the dynamics of the system under study.

essence, the problem is that it is not always obvious what precisely the system is that we are studying. The basic theory of alternative attractors describes the response of the system to *external conditions*. We consider the system to have alternative attractors if it can be in more than one stable state for the same value of such an external condition. It is important that this condition is really external, in the sense that it is not an interactive part of the system (figure 6.1). Reasonable examples are the effect of nutrient inflow on a lake or the effect of the Earth's orbit on solar radiation in the Sahara.

In ecology, a major misconception is the idea that all abiotic conditions are "external." This neglect of the view of an ecosystem as an interactive biotic–abiotic entity[1] has led to the idea that situations in which local abiotic conditions differ between the alternative states cannot be considered "proper" alternative stable states.[2] This would exclude the examples of plants promoting local water availability and many other important cases of alternative stable states known today.[3] Perhaps one of the underlying problems is that it is still counterintuitive to many that physical conditions may be determined to a large extent by biota. For instance, when our group wrote about the interaction between vegetation and climate in the Sahara–Sahel region in a review,[3] a well-known Dutch scientist sent a letter to a local science magazine claiming that we were confusing cause and effect here. Climate drives vegetation, not vice versa, was his reasoning. The concept of a strong two-way interaction between environment and biota is clearly not accepted by everyone. However, even if one accepts that the environment can be altered by biota, some insist that interactive change of biota and local environmental conditions implies that one cannot speak of alternative stable states. For instance, some ecologists objected that the alternative stable states in shallow lakes do not qualify as such, because the clarity changes and different resulting underwater light conditions for the plants would imply that we cannot say that we have the two states under the same "conditions." The misconception again is that *abiotic* conditions are not part of the interactive system of which we study the stability.

It is also important to note that the requirement that the external condition under study should not be affected by the system becomes less important if change in this condition is very slow relative to the rates of change in the system we study. We may then treat the slow variable as an independent control parameter for simplicity.[4] Note that it is the relative difference in rates that matters, not the absolute rates. For instance, lake plankton dynamics are fast relative to change in fish biomass,[5] but the collapse of Sahara vegetation (even if it took more than a century) was fast compared to the driving change in the Earth's orbit.[6] As shown earlier (section 3.1), in some systems, fast and slow components affect each other mutually in a way that leads to cycles. Recurrent pest outbreaks are a well-known example. Also, such slow–fast cycles can be understood by considering the processes at the slow and the fast timescales separately.

6.2 Basins of Attraction

Stability landscapes are perhaps the icon of the theory of alternative stable states. They are useful for transmitting the central concepts in this field. Yet they remain models, and consequently do not capture everything.

One of the main caveats in this interpretation is the fact that the image of a stability landscape suggests that the ball will gain *momentum* as it rolls down the hill and therefore may, for instance, overshoot

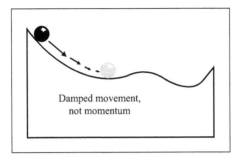

FIGURE 6.2. Intuitively, one might imagine a rolling ball to overshoot a local equilibrium. However, the speed of movement at any moment should be thought of as proportional to the steepness of the slope. Thus, there is no momentum gained by rolling balls in stability landscapes. Instead, they should be imagined as moving through a viscous fluid.

a small valley. However, this part of the intuitive message is wrong. The landscapes can be computed from a model in such a way that the slope represents the local rate of change.[7] But there is no momentum. Thus, one has to imagine the movements to be strongly damped (figure 6.2), as if the balls were actually rolling through a very viscous fluid.

Another important limitation of the classic stability landscapes is that they are *one-dimensional* in the sense that in addition to the external conditioning factor, they show only one variable ("dimension") of the system. Obviously, there are always many variables involved in real societal or environmental problems. Therefore, stability landscapes can represent only a fraction of the truth. "Real" attraction basins (the valleys) should be structures in many dimensions. This is hard to imagine, but to see some essential consequences of adding dimensions, it suffices to plot the boundary of the attraction basin in two dimensions (figure 6.3). A rather obvious observation is that in order to escape from the basin (that is, cross the border), it may be more effective to give a push to some variables than to push other ones (figure 6.3a). Perhaps less straightforward is the observation that pushing two variables simultaneously can sometimes be much more effective than pushing either of them alone very strongly (figure 6.3b). (Similarly, there can be situations in which pushing two variables simultaneously is less effective than pushing only one.)

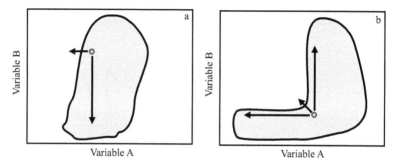

FIGURE 6.3. Hypothetical attraction basins (shaded) plotted in two dimensions reveal aspects of resilience that we cannot grasp from simple stability landscapes. Panel (a): The system may be flipped out of the attraction basin much more easily by pushing one variable (*A*) than by pushing the other (*B*). Also, pushing two variables simultaneously can sometimes be much more effective (panel b) than pushing either variable alone.

A third limitation of stability landscapes is that they do not help to understand how *unstable attractors,* such as limit cycles or strange attractors, can exist. One could think of a limit cycle as a circular channel, but we cannot see how the movement is maintained in the supposedly "heavily damped" system, and why it speeds up and slows down on certain sections.

A last limitation to keep in mind with respect to the interpretation of stability landscapes is the suggested *constancy* of the landscape. Most work stresses that stability landscapes may change gradually as the result of a slow change in the conditions. If such slow change causes a basin of attraction to shrink, resilience of the corresponding attractor is reduced, and stochastic shocks that displace the system (for instance, by killing part of some populations) may more easily flip the system into an alternative basin of attraction. However, the divide between slow change in conditions and fast stochastic shocks that occasionally push the system away from equilibrium is really somewhat artificial. In practice, conditions are subject not only to slow trends but also to stochastic fluctuations. This implies that one might imagine stability landscapes as vibrating and wobbling structures (figure 6.4). Again, looking deeper into that subject, timescales become important. If such fluctuations in conditions (affecting the

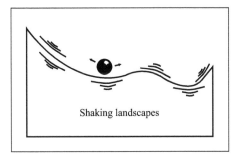

Shaking landscapes

FIGURE 6.4. Stochastic variation in conditions implies that the stability landscape should actually be viewed as a wobbling structure. This is in addition to the slow trends in shape that affect resilience and the stochastic shocks due to disturbance of the system that incidentally displace the ball.

landscape shape) are fast relative to the response time of the system, they are essentially filtered out. It is like exposing an elephant to a millisecond of extremely low temperatures. Even if the elephant cannot live at such temperatures permanently, a very short temperature drop will not even be noticed. In contrast, variations in conditions that are slow enough to let the system track the "moving target" that the attractor becomes will affect its dynamics.

6.3 Resilience

Resilience is quite difficult to define in a way that is both precise and useful. In a general sense most would agree on definition 1: *the capacity of a system to recover upon disturbance.* Although this is fine for everyday intuitive use, we have to be more specific if want to be able to quantify resilience. Probably the most widely used specific indicator is definition 2: *the speed with which the system recovers upon disturbance.* However, this does not capture the aspect of robustness against being tipped into an alternative basin of attraction. To deal better with this issue, definition 3 uses Holling's view of resilience[8]: *the magnitude of disturbance that a system can tolerate before it shifts into a different state (stability domain) with different controls on structure and function.* This corresponds to interpreting resilience as the width of the basin of attraction in stability landscapes. Holling coined the term "ecological resilience" for definition 3 to distinguish it from the recovery-rate definition 2, which he calls "engineering resilience."[9] Since it applies to many more systems than ecosystems and Holling deserves the credit for pointing out the importance of this aspect, one should perhaps use the term *Holling resilience* for the ability of the system to absorb perturbations without shifting to an alternative basin of attraction. However, I will simply refer to this concept as *resilience* in this book.

Both resilience and return time can be measured quite well in models, although one has to become specific about some aspects.[10,11] For instance, one has to define the disturbance precisely. In a system with more than one variable, it can make a lot of difference which variables we disturb. Some variables may be slower to recover than others, and the distance to the border of the basin of attraction depends on the direction of disturbance (cf. figure 6.3). If one wants to measure return time, another problem is that absolute recovery rates tend to zero as the system approaches equilibrium. Therefore, not only the magnitude of the perturbation is important, but also the point at which one considers the system (approaching the equilibrium asymptotically) to be "recovered." Last, some systems are simply much slower than others (for example, forests versus plankton communities), and one may wonder how interesting the return time measure is to compare such widely different systems. The bottom line is that it is difficult to obtain any absolute indicator of resilience or return time that is interpretable in a simple way. Therefore, instead of measuring resilience or return time in absolute terms, it is usually more meaningful to look at relative change as conditions are altered. Note that although the rate of return to equilibrium upon a small perturbation (definition 2) and the maximum perturbation that can be tolerated without crossing the border of the attraction basin (definition 3) appear to be two very different things, they are actually well correlated in most models.[11] This implies that rates of recovery upon small experimental or natural perturbations may be interpreted as indicators of resilience (see also section 15.2).

While measuring resilience in models is not easy, the greater challenge is of course to probe resilience in real systems. One of the fundamental difficulties here is that complex natural systems and societies usually do not recover to precisely the same state after a perturbation. Unlike in most models, there are countless variables, and many of those variables may replace each other without implying much of a fundamental change in the system (section 4.2). Obviously, it would be quite a coincidence if such complex systems would really recover to the same state (or dynamical regime). Ecosystems almost always recover upon disturbance to a state with a somewhat different species distribution. Similarly, societies recover upon perturbations to a state with different relationships, livelihoods, institutions, and companies. To deal with this we need a more liberal definition of resilience, as in definition 4: *the capacity of a system to absorb disturbance and reorganize while undergoing change so as to still retain essentially the same function, structure, identity, and feedbacks.*[12] In ecosystems, species richness will contribute to such resilience, as different species may take over a certain functional role (for example, grazing control of algae) if a disturbance hits more-sensitive species in the same functional group (see section 5.3). The interpretation of "essentially the same function, structure, identity, and feedbacks" will differ from case to case. The interesting question for a disturbed coral reef is not which species will be precisely present in the community, but rather whether it recovers to a coral-dominated state rather than an algae-dominated condition. For a lake, we may be interested in whether the lake recovers to a clear state or slips into a contrasting turbid state rather than which planktonic species will be present.

6.4 Adaptive Capacity

Societies differ from ecosystems in that processes such as innovation may largely determine how well they may regain "essentially the same function, structure, identity, and feedbacks" upon disturbances. Hence, social resilience may depend mostly on *adaptive capacity* and reflects the degree to which a system is capable of reorganization, learning, and adaptation.[10] Such adaptive capacity is essential to ensure the persistence of many social systems. In ecosystems, adaptive capacity is determined largely by the (response) diversity of species. However, it is becoming clear that rapid evolution may well take a role similar to that of learning and innovation in social systems when it comes to the response of ecosystems to global change. I will come back to the issue of adaptive capacity in more detail at the end of this book.

6.5 Critical Transitions

Last but not least, let us come back to the core issue of this book. What exactly is a *critical transition*? If the book were just about models, it would have been sufficient to use the dynamical systems terminology and talk about shifts between alternative attractors that happen in the vicinity of catastrophic bifurcations. However, it will be obvious by now that we can never really map the large complex systems of nature and society on the mirror world of math. To avoid confusion, I therefore prefer to avoid the use of those precisely defined concepts of dynamical systems theory. On the empirical end of the spectrum, we have the term *regime shift*, which is used widely used to describe a sudden jump from one dynamic regime to another one.[13] As discussed in more detail later (section 14.1), such regime shifts are phenomena that do not necessarily imply the existence of alternative basins of attraction. For instance, in ecosystems, they can also happen as the result of a large stepwise change in the environmental conditions. Likewise, in socioeconomic systems, they may result from a sharp external change such as an important political decision. I reserve the term *critical transition* for the subclass of regime shifts that in models would correspond to shifts between alternative attractors. Those are the transitions in which a positive feedback pushes a runaway change to a contrasting state once a threshold is passed.

6.5 Synthesis

In conclusion, there are essential nuances if one is to make practical use of each of the five key concepts that are central to this book: *alternative stable states*, *stability basins*, *resilience*, *adaptive capacity*, and *critical transitions*. The concept of *alternative stable states* applied to the real world refers to the phenomenon that under the same external conditions, a system can maintain itself in two or more contrasting dynamical regimes. External conditions in this context are factors that are not notably affected by the system under study or that have very slow dynamics compared to those of the system.

The fact that systems can have alternative basins of attraction and some important implications of this can be illustrated in an intuitive way by means of *stability landscapes*. However, such stability landscapes should not be taken too literally as they do not capture the damped nature of change (that is, the absence of momentum), the multidimensional nature of systems, the mechanics of cycles and strange attractors, and stochastic fluctuations in conditions.

Resilience is the capacity of a system to recover upon disturbance. The speed of recovery is often used as an indicator. More interesting in the context of this book is the magnitude of disturbance that a system can tolerate before it shifts into a different state. This resilience tends to be correlated to the rate of recovery. In any case, absolute measures of resilience are difficult to interpret, and change in resilience in response to altered conditions is more informative to study.

Adaptive capacity is the degree to which a system is capable of reorganization, learning, and adaptation. It is an essential feature of an alternative definition of resilience that is less precise but often more realistic—namely, the capacity of a system to absorb disturbance and reorganize while undergoing change so as to still retain essentially the same function, structure, identity, and feedbacks. In social systems, adaptive capacity depends not only on diversity (as in ecosystems) but also on processes such as learning and innovation.

Critical transitions are sharp shifts in systems driven by runaway change toward a contrasting alternative state once a threshold is exceeded. This implies that they are a subset of regime shifts, which includes the broader phenomenon of shifts between persistent contrasting states.

Part II
CASE STUDIES

The previous chapters have been written from the perspective of dynamical systems theory. Starting with the general rules of the game, we explored what had to be added to the basics in order to cover the rich dynamics observed in real systems. Now we approach the problem from the other, perhaps more natural side. As we did briefly in the introductory chapter, we will look at surprising dynamics in nature and society, but we now explore how we might map them onto the mirror world of dynamical systems theory. As you will see, the link can be made quite well from some problems. In other cases, we just note a similarity between patterns in the real world and predictions from dynamical systems theory, without being sure that the same mechanisms are really at work. Such hints point at potentially rewarding fields for further research, because if we are able to pin down the mechanisms at work, this may eventually open up the possibility of predicting, preventing, or catalyzing big shifts in nature and society.

I have written part II in such a way that you can use it as a starting point for reading the book, depending on your taste. Starting here, you come from the practical side and can then dig into the more systematic treatment of the theoretical aspects when you are ready. This dual-entry structure makes a little bit of overlap inevitable. However, in most instances, I refer to other chapters when it comes to the details.

CHAPTER 7
Lakes

A few decades after Charles Darwin published *The Origin of Species*, the American scientist Stephen A. Forbes wrote an influential article titled "The Lake as a Microcosm."[1] The article basically describes the dynamics of the ecosystem of a floodplain lake upon retreat of the river water. However, he characterizes the image of the intrinsic web of interactions involved in the struggle for survival so well that it inspired others to look at ecosystems this way. Therefore, his contribution is often considered the start of ecology as a scientific discipline. In fact, Forbes was arguing that lakes may be seen as little worlds that reflect what is going on in the big world, a subtle twist on the view that the Greeks had of our minds as Mikros-Kosmos reflecting the entire world outside. Forbes did not merely think that lakes reflected how other ecosystems might work; he also explicitly suggested an analogy to the way in which human societies work.

As you may guess, I feel somehow the same about what we can learn from lakes. Most of my work has been on lake ecosystems. But over the years, I have interacted with scientists in fields ranging from dryland ecology to climatology, oceanography, economy, and sociology in an attempt to grasp the forces that govern *social–ecological systems*. Time after time, in some way or another, mechanisms analogous to what I had seen in lakes appeared to operate. Often, analogies appear on high abstraction levels only, but similarities are often strong enough to allow them to be characterized by similar mathematical equations and graphs. That should be no surprise. After all, they are all dynamical systems, and eventually they obey the generic rules that apply to any dynamical system. In fact, any system can serve as an analogy to any other system in some aspects. What makes lakes special and inspiring is that they are extensive enough to harbor many scales of complexity but still just the size that can be well overseen and even experimentally probed. My familiarity with the ecology of shallow lakes and the richness of the mechanisms found in lake studies are the reason that I will dwell a bit longer on what was found in these ecosystems before I proceed to case studies from other natural systems and societies.

7.1 Transparency of Shallow Lakes

The water of a shallow lake or pond may offer a fascinating view of submerged plants, fishes darting off, and small animals moving busily around. Often, however, the water is murky, troubled by blooming algae and suspended sediment particles hiding whatever is going on below. The fact that the latter situation has become the most common one in most regions is due to high nutrient (especially phosphorus and nitrogen) loading. The simple version of the story is that nutrient enrichment stimulates the growth of phytoplankton to the point that dense populations of these microscopic suspended algae make the water turbid. However, restoration efforts aimed at reducing the nutrient content have been puzzlingly unsuccessful in such lakes. In contrast, an additional push by reducing the fish density for a brief period has brought many of them back to a permanently clear condition (figure 7.1). This has triggered some decades of intensive research demonstrating that the clear and the turbid situation represent strongly contrasting community states, both of which have stabilizing feedback mechanisms.

An Introduction

Traditionally, *limnology* (the study of inland waters) is mostly concerned with lakes that stratify in summer. Thermal stratification largely isolates the upper water layers (*epilimnion*) from the colder deep water (*hypolimnion*) and from interaction with the sediment during the summer. The impact of rooted submerged vegetation (*macrophytes*)

FIGURE 7.1. Large-scale experiments have demonstrated that temporary reduction of the fish density can tip some turbid lakes into a stable clear state.

on the community is relatively small in such lakes, as plant growth is restricted to a narrow marginal zone. In contrast, shallow lakes can be largely colonized by macrophytes and do not stratify for long periods in summer. The average depth of these lakes is usually less than 3 meters, but their surface area ranges from less than a hectare to over 100 km^2. The intense sediment–water interaction and the potentially large impact of aquatic vegetation makes the functioning of shallow lakes different from that of their deep counterparts.

In many regions, shallow lakes are more abundant than deep ones. Floodplains of rivers are often seeded with such lakes, and numerous shallow lakes are found at the edge of the ice cover during the last glaciation period. Also, human activities such as digging for peat, sand, gravel, or clay have produced considerable numbers of shallow lakes and ponds. The term *wetlands* is often used to refer to shallow lakes and adjacent marshy land. Such habitats are notoriously rich in wildlife. In densely populated areas, even small lakes can be very important from a recreational point of view. Fishing, swimming, boating, and bird-watching attract a large public.

The pristine state of the majority of shallow lakes is probably one of clear water and a rich aquatic vegetation. As mentioned earlier, nutrient loading has changed this situation in many cases. The lakes have shifted from clear to turbid, and with the increase in turbidity, submerged plants have largely disappeared. The sequence of changes during eutrophication is rarely documented well, but some elements are agreed upon by most workers in the field.[2] Shallow lakes with a low nutrient content usually have vegetation dominated by relatively small plants. With increased nutrient loading, the biomass of aquatic macrophytes increases, and plants that fill the entire water column or concentrate much of their biomass in the upper water layer become dominant. Such dense weed beds are often experienced as a nuisance by the fishing and boating public. When weed control programs eradicate the vegetation, turbidity in shallow lakes tends to increase strongly because of algal blooms and wind resuspension of the sediment. Also, when vegetation is not controlled explicitly, further eutrophication of vegetated lakes can lead to a gradual increase of phytoplankton biomass and of the greenish *periphyton* layer that covers the plants. Shading by these organisms ultimately leads to a collapse of the vegetation because of light limitation.

Mechanisms

Restoration of nonvegetated turbid shallow lakes to the clear vegetated state is notoriously difficult. Reduction of the nutrient loading may have little effect, as during the period of eutrophication, a large amount of phosphorus has often been adsorbed by the sediment. When the loading is reduced and its concentration in the water drops, phosphorus release from the sediment becomes an important nutrient source for phytoplankton. Thus, a reduction of the external loading is often compensated by *internal loading*, delaying the response of the lakewater concentration to the reduction of external loading.

However, internal loading is not the only reason why restoration of turbid shallow lakes is difficult. With the disappearance of aquatic vegetation, the structure of the shallow lake community changes dramatically (figure 7.2). Invertebrates that are associated with vegetation disappear, and with these animals, the birds and fishes that feed on them or on the plants. Also, vegetation provides an important refuge against predation for many animals, and hence its disappearance causes crucial shifts in many predator–prey relationships. Large zooplankton use vegetation as a daytime refuge against fish predation.

FIGURE 7.2. Schematic representation of a shallow lake in a vegetation-dominated clear state (top panel) and in a turbid phytoplankton-dominated state (bottom panel) in which submerged vegetation is largely absent and fish and waves stir up the sediments.

In vegetated lakes, they can contribute significantly to the control of phytoplankton biomass. In the absence of vegetation, their numbers are strongly reduced. The lack of waterfleas (*Daphnia*) and the increased nutrient availability allows phytoplankton biomass to be higher in the absence of vegetation. In addition, wave resuspension of the unprotected sediment can cause a considerable additional turbidity once the vegetation has vanished. The fish community of unvegetated lakes becomes dominated by species that forage on sediment-dwelling small animals such as worms and various insect larvae. Their activity promotes the nutrient flux from the sediment into the water and causes an extra resuspension of sediment particles, contributing to the already high turbidity.

Return of submerged plants in this situation is unlikely, in part because their absence has allowed a further increase in turbidity, but also because the frequent disturbance of sediment by wind and by fish that forage on midge larvae and other animals that live in the sediment hampers resettlement. Ecological feedback mechanisms are thus an important reason why restoration of the vegetated clear water state is difficult. In many cases, nutrient reduction alone may be insufficient to restore the clear state in shallow lakes. Additional measures, however, such as removal of part of the fish stock and alteration of the water level, have been successfully used as a way to break the feedback that keeps such lakes turbid.

We will now look more precisely at how ecological mechanisms can cause multiplicity of stable states in shallow lakes. The interaction between submerged plants and turbidity is thought to be a major driving mechanism explaining the overall response of shallow lake ecosystems to nutrient loading and restoration efforts. Vegetation tends to enhance water clarity, but lack of light in turbid water is one of the main problems for submerged plants in eutrophic lakes. This implies a positive feedback in the development of submerged vegetation: once the plants grow, the water clears up, and they grow even better. Figure 7.3 summarizes the main mechanisms involved. A simple way of evaluating the overall effect of the depicted interactions is to multiply the signs along the way of a path through the scheme. This exercise shows that through all depicted routes, turbidity enhances turbidity, and vegetation enhances vegetation.

The existence of stabilizing mechanisms that tend to keep the system in either a vegetation-dominated state or a phytoplankton-dominated state suggests the potential for alternative equilibria. However, in mathematical models, alternative equilibria usually occur for limited ranges of parameter settings only, and likewise, real systems will normally have these properties only for a limited set of conditions. Indeed, the hypothesized stabilization of the vegetated state seems unlikely in deep lakes where the narrow littoral zone that can be vegetated has a less-dramatic impact on turbidity than in shallow lakes that can be entirely vegetated. Also in shallow lakes, the existence

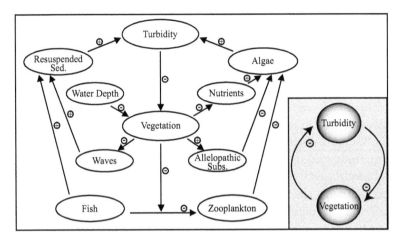

FIGURE 7.3. Feedbacks that may cause a vegetation-dominated state and a turbid state to be alternative equilibria. The qualitative effect of each route in the diagram can be computed by multiplying the signs along the way (so two minuses result in a plus, and so on). This shows that vegetation has a positive effect on itself through different feedback loops, but the same holds for turbidity. Thus, both the vegetated and the turbid state are self-reinforcing (inset). (Adjusted from reference 3.)

of alternative stable states will be limited to an intermediate range of nutrient levels, as nutrient-poor lakes are rarely turbid and very high nutrient loading usually excludes vegetation dominance. Therefore, the demonstration of stabilizing mechanisms per se is not sufficient to infer that a lake may have alternative stable states (a point discussed in a more general context in section 14.3).

MODELS

A simple graphical model (figure 2.9) presented earlier explains in an intuitive way how the effect of nutrient loading on the potential for alternative equilibria can be understood. Note that in this figure, the horizontal dashed line representing the critical turbidity at which vegetation disappears will depend on how deep the water is. Deeper water implies that light limitation for the submerged plants occurs already at lower turbidity. Interpreted this way, the model explains why a change in water level may also induce a shift to an alternative state. Indeed, years with high water level have been related to a shift to the turbid vegetationless state in many lakes, and conversely, low water levels may induce a shift to the vegetated clear state. Clearly, this graphical model is a crude simplification of the interaction between vegetation and turbidity. Especially the assumption that all plants vanish at a single critical turbidity is a gross oversimplification.

To obtain a more sophisticated view, we have to consider the effect of vegetation on turbidity as well as the effect of turbidity on vegetation in more detail. I briefly explain such a model here to make our case study section more self-contained (figure 7.4); for a mathematical model, see the appendix, section A.12). Other graphical models explaining mechanisms for alternative stable states and catastrophic shifts can be found in section 2.2.

Although the effect of vegetation on turbidity is related to a complex set of mechanisms (figure 7.3), the overall result is fairly predictable, as shown experimentally and in field patterns (figure 7.4). The effect of turbidity on the vegetation cover in a lake depends largely on the depth profile of that lake. Patterns across many lakes show that the depth to which vegetation can grow is related in a predictable way to the turbidity of the water. With increasing turbidity, vegetation disappears first from the deepest sites because of light limitation. This implies that if lakes have a frying pan shape with the same water depth everywhere, vegetation for the most part disappears at a single critical turbidity. However, in lakes that have a more concave shape, the resulting loss of vegetation cover with turbidity will follow a roughly sigmoidal path (figure 7.4b). We can thus plot how the equilibrium turbidity is affected by vegetation cover (7.4a) and how equilibrium vegetation cover is affected by turbidity (7.4b). If we plot those two equilibrium lines together (thus reflecting the axes in panel a), the intersection points represent situations in which both vegetation and turbidity are in equilibrium (7.4c). It can be seen from the arrows that indicate the direction of change when the system is out of equilibrium that the middle intersection point represents an unstable *saddle point* that attracts trajectories from two directions but repels them in two other directions. The other two intersection points are stable *nodes* that attract trajectories from all directions.

Despite its simplicity, this graphical model highlights some key features of the stability properties of shallow lakes that are also found in

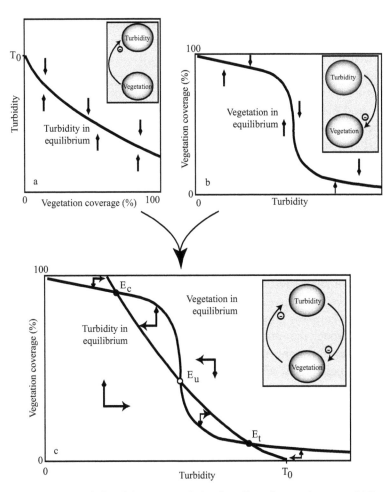

FIGURE 7.4. For shallow lakes, we can derive the effect of vegetation on turbidity (a) and the effect of turbidity on vegetation cover (b). Combining this information, we obtain a model that shows the equilibria (solid dots for stable; open dots for unstable) resulting from the dynamic interaction between turbidity and vegetation (c). Arrows indicate in which direction both variables will change outside the equilibria. T_0 is the turbidity of the water in the absence of vegetation and will be higher if the nutrient load to the lake is higher.

realistic individual-based spatially explicit simulation models.[4,5] First, hysteresis in response to change in nutrient loading can be easily deduced. As nutrient level will affect turbidity, it will shift the entire turbidity curve (figure 7.4a), causing the intersection points to shift too (figure 7.4c) and eventually merge and disappear in bifurcation points (F_1 and F_2 in figure 7.5). The same will happen if one were to change the water depth in this imaginary lake, as this will affect the critical turbidity around which the vegetation curve (figures 7.4b and c) drops.

It can also be seen that it is possible to have multiple intersections (and hence multiple equilibria) only if the drop in the vegetation curve is steep enough to exceed the slope in the turbidity curve (figure 7.4c). The latter is vertical if vegetation does not affect turbidity, allowing no multiple intersections. Thus, the stronger the effect of vegetation on turbidity, and the steeper the drop in vegetation at a critical turbidity, the higher are the chances to have a substantial range of nutrient loadings with alternative stable states (and thus, *hysteresis*). One variable that affects both curves is the depth of a lake. In deeper lakes, plants are likely to have less impact on clarity than in shallow lakes where they reach through the entire water column. Also, the usually gradual slopes in deeper lakes imply that plant cover declines gradually (starting with the deepest places, where light becomes limiting first) rather than abruptly with increasing turbidity. As a result, multiple intersections of the equilibrium curves are less likely as lakes are deeper, and there may be no hysteresis in the response to nutrient loading (figure 7.5b). This illustrates the general phenomenon that systems with alternative attractors that respond to environmental change through hysteresis and catastrophic transitions may actually turn into systems that respond smoothly to change in conditions (figure 7.5c) as critical parameters (in this case depth) change. It can also be seen that there will not be a single critical nutrient level at which a lake shifts from clear to turbid, as this will depend on depth (and many other aspects). Again, this illustrates a general rule: critical levels of one factor will usually depend on the value of other factors. Such generic rules are discussed in more detail in chapter 2. In the next sections, we will illustrate these and other general "laws" further using the shallow lakes as a case study.

If you are interested in playing around with this lake model on the computer, the appendix (section A.12) gives equations. The behavior of shallow lakes has been explored by means of various more elabo-

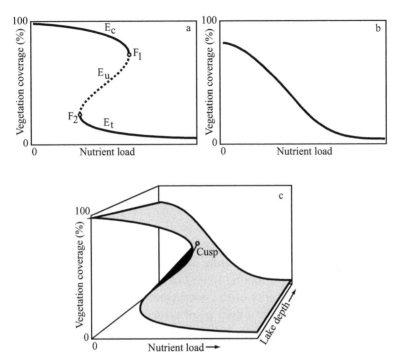

FIGURE 7.5. Response of submerged vegetation to nutrient load in shallow lakes, as derived from the graphical model presented in figure 7.4. Hysteresis can be strong (panel a) or absent (panel b) depending on the shape of the equilibrium curves. In reality, all kinds of responses will exist (c) varying from smooth (b) to hysteretic (a), and the latter seems more likely in shallower lakes. The transition from hysteretic to smooth happens in a so-called *cusp point*. This is a co-dimension-2 bifurcation point where fold bifurcation points F_1 and F_2 merge and disappear.

rate and realistic models. One line of work has focused into more detail on the aquatic vegetation and spatial heterogeneity.[4,5] These models describe the seasonal life cycle and the growth of individual plants in interaction with their dimensional environment. Another line of models explicitly includes population dynamics of vegetation, phytoplankton, zooplankton, and fish, as well as nutrient dynamics resuspension and other processes.[6,7] All of these models appear to have alternative stable states over a range of conditions.

Still, no model can cover all mechanisms that may be of importance. For instance, turbidity is probably not the only reason why submerged plants are unable to colonize some shallow lakes. Herbivory by birds may keep sparse vegetation from developing further. This mechanism can hamper the recovery of vegetation in lakes where transparency has increased in response to nutrient reduction or reduction of the fish stock. Although herbivores, such as coot, are usually not abundant enough to wipe out existing vegetation, even low consumer densities may be sufficient to prevent a sparse vegetation from recovering.[8] Differences in the fish communities that develop in vegetated versus unvegetated lakes are also thought to play a major role in maintaining existing conditions. Large bottom-feeding fish such as bream and carp tend to dominate unvegetated lakes, and settlement of plants is unlikely in systems where the sediment is continuously disturbed by their "weeding" activity. Vegetated lakes, on the contrary, have few of such bottom-feeding fish and instead develop relatively high densities of fish-eating fish such as perch or pike that may control abundance of young fish and thus enhance survival of zooplankton, thus promoting water clarity and stabilizing the vegetated state.[9] The stability of the nonvegetated state may also be enhanced by the fact that initial plant settlement can be difficult in wind-exposed lakes if the sediment is soft and unstable. It is hard to envision small plants settling when the upper centimeters of the sediment are involved in frequent resuspension. Established vegetation stands will dampen wave action and allow consolidation of the sediment, facilitating further colonization. In fact, this mechanism could lead to alternative stable states even if the vegetation–turbidity feedback did not exist.[10] In conclusion, the vegetation–turbidity feedback tends to cause alternative stable states in shallow lakes but other mechanisms may well contribute to the hysteresis.

Clearly, even the elaborate models of shallow lake dynamics capture only part of what is going on. All of the shallow lake ecosystem models so far appear to have alternative stable states over a range of conditions,[4,5,7] and in view of the difference in approaches, this supports the idea that phenomenon may be common in these ecosystems. As Levins[11] once phrased it, one is more likely to accept something as the truth when it emerges "as the intersection of independent lies." Although "lies" may sound a bit too harsh for the models involved, all of these models clearly have their limitations. Also they are not really independent, as they build on a common set of ideas circulating in the shallow lakes literature.

EMPIRICAL EVIDENCE

The strength of the shallow lake case really stems from the fact that we have not only good mechanistic insights and a large range of models of different foci and complexity, but also convincing empirical evidence derived from observations of natural dynamics in hundreds of lakes as well as the response of such ecosystems to whole-lake experimental manipulations. To start with natural patterns, shallow lakes tend to be either clear or turbid, and rarely something in between. This impression is not an artifact of our natural tendency to simple black-and-white, good-or-bad classification. It is supported by statistical analyses of the distribution of states in large populations of shallow lakes.[12,13] For example, of the 215 shallow lakes sampled in the floodplains of the lower river Rhine, more than one third was practically devoid of vegetation, whereas most of the remaining ones were densely vegetated (figure 7.6a). Note that, as explained in section 14.1, such bimodality is suggestive but no proof of alternative stable states.

Regime shifts are a similar kind of pattern that might or might not correspond to catastrophic transitions between alternative attractors (section 14.1) but that are suggestive nonetheless. Time-series of various decades are required to decide whether such a shift indeed represents a sharp transition between persistent contrasting states. Such time-series that happen to include a state shift are relatively rare, but they do exist. The history of the Swedish lake Krankesjön[14] is an example (figure 7.6b). Systematic waterfowl counts allowed the reconstruction of one century of dynamics. Over this time span, the lake has been clear for two periods and turbid over two other periods. Transitions may take 3 to 4 years to complete in this lake but can be much more rapid in smaller lakes.[15]

Even less common are time-series that cover a complete hysteresis loop over a period of increasing and subsequently decreasing nutrient load. The Dutch lake Veluwe is one example (figure 7.6d). Data on loading are rare, and the plotted total-P content of the water is an imperfect indicator of the nutrient status. Nonetheless, it is interesting to see that this time-series indicates that the critical P-level for collapse is about two times higher than the critical P-level for recovery. Hysteresis of roughly this magnitude is independently predicted by several

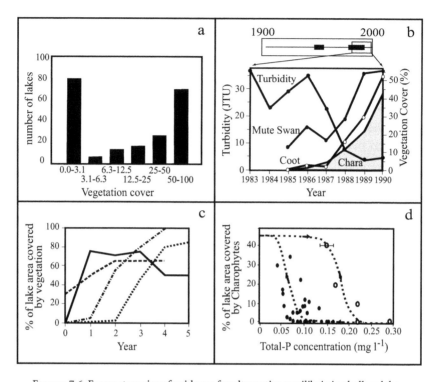

Figure 7.6. Four categories of evidence for alternative equilibria in shallow lakes: (a) bimodal distribution of vegetation abundance found in a set of 215 floodplain lakes (reference 11); (b) sharp regime shifts interrupt long periods of stable turbid (heavy line) or clear (light line) states in the Swedish lake Krankesjön (reference 13); (c) an experimental disturbance (fish reduction) can bring lakes into a long-lasting contrasting state (reference 13); and (d) long time-series suggest hysteresis in the response of the Dutch lake Veluwe to nutrients (reference 28).

mechanistic models[5,7] and has been deduced from time-series of another Dutch lake too.[16]

Although these observations are suggestive, more important evidence comes from experimental manipulations. Some cases consist of exceptional natural experiments such as the violent storm hitting the New Zealand lake Ellesmere in 1968.[17] The lake had a population of 80,000 black swans living on abundant aquatic vegetation. The storm that instantly killed 5,000 swans also wiped out much of the aquatic vegetation. The lake has shifted to a turbid state, with scarce vegetation and only 4% of the original swan population, from which it has not re-

covered since. Although several cases of permanent effects of such natural perturbations have been documented,[10] the most important class
of empirical evidence for alternative attractors in shallow lakes comes
from their response to human interventions. Many turbid shallow lakes
have received a shock treatment in an attempt to restore the clear state,
typically intensive fishing in winter to remove a good part of the fish, a
measure known as *biomanipulation*.[18] If more than about 80% of the
fish has been removed, this leads almost always to a great increase in
water clarity.[19] Fish biomass usually recovers quickly. But different species tend to become dominant, and part of the lakes that have been perturbed this way have remained in an apparently stable vegetated and
clear state for many years (for example, figure 7.6c). A particularly interesting case is that of the English Great Linford sand and gravel
pits.[20,21] Two distinct methods of gravel extraction were used in this
area: wet-digging and dry-digging. Wet-digging results in silt-laden
lakes, whereas in the case of dry-digging, clear water fills the lake when
pumping ceases. Remarkably, wet-dug lakes have typically remained
turbid, even though most of them have been left undisturbed for over
20 years. They have very little submerged vegetation, and the sediment
is flocculent and easily resuspended by waves that can turn the water
chocolate brown. Thus, difference in initial conditions has led to permanent alternative states, an important indicator of alternative attractors (see section 14.2). This impression has been confirmed by the fact
that experimental biomanipulation of one of the turbid lakes has led to
a stable clear state with a dense cover of submerged plants.[19]

FACTORS AFFECTING THE THRESHOLD

As eutrophication problems have been an important incentive for much
of the work on shallow lakes, one of the central questions has always
been what the critical nutrient level would be for maintaining a clear
water state. Although this seems a straightforward question, there are
some fundamental problems if one attempts to find the answer. First,
it is not so easy to assess the *nutrient level* of a lake. For instance, although phosphorus is clearly important, nitrogen may play an important role in some cases too. Also, when it comes to phosphorus, much
of what is available to the organisms in the long run is stored in the
sediment. How much of that is reflected in total-P concentrations in

the water column varies strongly with the presence of macrophytes and other biological factors. Thus, although we often use total-P as an indicator of the nutrient status of a lake, there are some caveats to that approach. Also, as pointed out earlier, there will not be a single critical nutrient level for maintaining clear water. Different lakes may vary widely in the nutrient level they can tolerate before they flip to a turbid condition. In principle, the list of factors that may influence the chance that a lake turns to the turbid state is almost endless. We have already seen the effect of water depth. A related factor that may have a large impact on the chances of a lake to be in a vegetation-dominated clear state is lake size. Although lake size tends to be correlated to lake depth, size seems to have a considerable effect by itself too. Small lakes appear to have a higher chance to be in a vegetated clear state. Although various factors may explain this correlation,[12] suppression of fish in smaller lakes may be a common factor. For instance, analysis of 215 shallow lakes situated in the Dutch floodplain of the lower river Rhine[12] revealed that the likelihood of a richly vegetated state was higher in smaller lakes (other factors such as depth being equal)[12] and that such small vegetated lakes also supported low densities of benthivorous bream (*Abramis brama*),[22] the key fish species promoting the opposite poorly vegetated, turbid state in shallow Dutch lakes.[10] Similarly, data from 796 Danish lakes and ponds[23] showed lower fish biomass and higher macrophyte coverage in smaller lakes and ponds despite a generally higher phosphorus content in the lakewater related to a generally higher share of cultivated fields in the catchment.[23] Thus, the results from both the Dutch and Danish lakes suggest that small lakes are more likely to be fishless, which increases the likelihood of the clear water state, even at quite high nutrient concentrations.

A major unresolved question is what would be the impact of climate on the chances that shallow lakes fall in a turbid state. Although we have little information about this, there are good reasons to expect a large climatic impact on these ecosystems. We know, for instance, that warmer conditions have a large effect on the *trophic cascade* from fish to phytoplankton. In temperate conditions, most fish reproduce only once a year, leaving a period in spring in which there are few small (juvenile) fish, allowing large zooplankton to become abundant and filter the water clear of phytoplankton.[24] In contrast, top-down control of zooplankton by fish is very strong all year round in warmer lakes at low latitudes because fish are abundant and reproduce continuously in such (sub)tropical lakes.[25] The difference in foodweb structure implies that biomanipulation as used to shift temperate lakes to a clear state seems less easy to apply to (sub)tropical lakes.[26]

One could expect that climatic warming might promote the turbid state in temperate lakes. On the other hand, numerous field studies in temperate lakes suggest positive effects of warming on aquatic vegetation performance,[27] which would push the other way. Importantly, climatic warming will also affect lake ecosystems through changes in hydrology and nutrient load to lakes, but findings so far are rather contradictary.[28] An interesting aspect of climatic impact is that systems with alternative stable states, even brief climatic extremes, may induce a shift to another state in which the system subsequently remains for a long time.[29] Indeed, there are indications that shallow lakes may be affected by climatic extremes in this way.[10] As mentioned earlier, heavy storms have induced shifts, and droughts can cause an opposite shift.

7.2 Dynamics

As described in section 5.1, the concepts of the equilibrium of nature or a stable state represent highly simplified images of the eternal turmoil that characterizes any real ecological situation. Such turmoil is due partly to fluctuations in external conditions, but can also be generated in part by internal mechanisms that would cause the system to display cyclic or chaotic behavior even under constant external conditions. To illustrate this, we look at (1) how lakes behave in situations with heavy environmental fluctuations, (2) how intrinsic dynamics can cause long-term cycles, and (3) how seasonal cycles interfere with fast intrinsic dynamics in plankton to cause complex patterns when we examine the ecosystem on a more detailed scale.

LAKES IN FLUCTUATING ENVIRONMENTS

We have seen that lakes can be flipped into another stable state by occasional disturbances such as fish removal or extremes in the water level. However, there are also situations in which disturbances are the rule rather than the exception, and it becomes more difficult to think of stable states. Such high-disturbance regimes are found, for instance, in cold climates where many lakes may freeze to the bottom and winter fish kills are common or in places where desiccation of shallow lakes in summer is a frequent event.[30] Under such conditions, lakes may be far from any equilibrium most of the time. Although the concept of alternative stable states is clearly too simple to catch the essence of the dynamics of lakes under such severe disturbance regimes, it may still help in seeing the big picture behind the dynamics.

For instance, in models, it can be shown that the time that systems stay away from attractors can be boosted if there are points or trajectories where they tend to "hang around" even if they do not represent attractors (section 5.1). As model systems may linger for long periods around unstable points or cycles, one would suspect that this kind of behavior could well dominate dynamics in nature too. So far, few studies have looked for such prolonged transient dynamics. However, the tendency of some shallow lakes to stick to an unstable clear state for years before sliding into a stable turbid state may be an example of such behavior. Lake Zwemlust, for example stayed clear for about seven years upon fish removal, despite a very high nutrient concentration.[31] Similarly, numerous Dutch floodplain lakes are in a clear state most of the time, despite high nutrient levels.[15] A likely explanation is that occasional droughts result in near desiccation of the lakes. This kills most of the fish and pushes the system into a clear situation that is not stable but that is almost so. This situation thus represents a *ghost* of a stable state, and the dynamics away from it are very slow (figure 7.7).

Multiyear Cycles

Although the long-term variations that can be observed in the state of lakes usually have a rather erratic character,[32] some lakes show remarkably regular oscillations between submerged plants and a turbid state (figure 7.8). The best-documented examples are the English lake Alderfen Broad[33] and the Dutch lake Botshol.[34] Both lakes cycle between the classic alternative states with a period of approximately 7

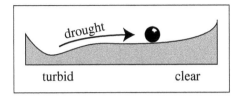

FIGURE 7.7. Hypothesized dynamics of shallow floodplain lakes. Occasional droughts push the system to an unstable clear state that represents the ghost of a stable state. Because droughts happen occasionally and the dynamics around such a ghost are very slow, the lake may be in a transient state most of the time. Note that the way to interpret such stability landscapes is that the movement of the ball is always damped. Imagine that the ball rolls through a heavily viscous fluid. (From reference 15.)

years. Model analyses suggest that such cyclic behavior may arise under particular conditions from an internal "time-bomb" mechanism[16]: if during the period of macrophyte dominance, phosphorus retention is high and dead organic material accumulates on the sediment, decomposition may eventually cause anoxic conditions at the sediment surface, allowing phosphorus release that may promote a sufficient increase in phytoplankton to cause a decline of the submerged macrophytes. If subsequently the turbid state allows sufficient decomposition of organic material and/or loss of phosphorus from the sediments, the lake may shift back to a clear macrophyte-dominated state, and so forth.

While such an internal time-bomb mechanism may be important in driving cycles of critical transitions in shallow lakes, the oscillations of Lake Botshol show a remarkable synchrony to interannual oscillations in precipitation driven by the North Atlantic Oscillation (NAO). In this lake, rainy years result in substantially higher phosphorus loading, pushing the lake toward the turbid state.[35] Since other cyclic lakes do not have this synchrony with the NAO rhythm, it seems most likely that an intrinsic mechanism is the general cause, while in Lake Botshol, the climatic oscillator serves as a pacemaker to lock the phase of the cycles. Such phase locking of oscillating systems is quite common in nature (section 3.2). For instance, quasi-periodic variations in the Earth's orbit are the pacemaker for the glacial cycles (section 8.2), and as explained in the next section, the annual cycle of temperature and light serve to lock cycles in plankton to a regular seasonal pattern.

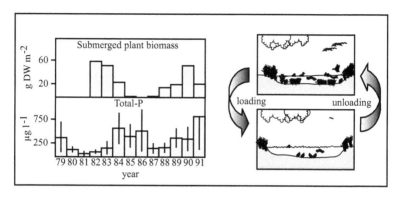

FIGURE 7.8. Shallow lakes may under some conditions shift between a vegetated and a turbid state in a regular cyclic way. This may be due to intrinsic differences in accumulation of phosphorus and organic matter in the two states (see text for explanation). (Data in the left panels are from the English lake Alderfen Broad; see Perrow et al., reference 33.)

SEASONAL DYNAMICS AND INTRINSIC CHAOS

One simple reason that temperate ecosystems can never have a stable state is the seasonal cycle. The effect of seasonality on dynamics depends on the longevity and life cycle of the organisms we consider. In temperate lakes, submerged vegetation often seems to disappear in winter. In reality, the plants are still there, but in the form of structures such as seeds, roots, and tubers that remain buried in the sediments. Thus, although it may seem that the ecosystem is reset in winter and has to start all over again to settle in a vegetation-dominated clear state, it really keeps a memory of the state of the past year. The underground structures of the plants have enough energy stored to ensure substantial regrowth. Note that there is a trade-off involved for the plants. During the growing season, plants have to start storing resources in overwintering structures; however, this comes at the cost of investment in leaves and stems. Therefore, the plant has to balance vigor of growth during summer and potential for regrowth next year. Clearly, both are important, and therefore, investing only in growth or only in overwintering structures will not work out. There must be an optimum strategy. Interestingly, it appears that different plants that are common in shallow lakes invest just the amount of energy in overwintering structures that simulation models suggest to be optimal for long-term survival of the population.[4] Thus, although some plants have short reproductive cycles to quickly respond to situations in which the lake bottom occasionally dries out, most are tuned to follow the regular seasonal cycles in temperate lakes.

The way seasonality drives the dynamics of fish communities is different. In spring, females typically produce huge amounts of eggs, and countless so-called young-of-the-year animals can be found along the shores some weeks later. They seek the shallow places where temperatures are high to stimulate their growth and where they are relatively safe from larger predatory fish. As the little fish grow, food becomes increasingly limiting, and many starve to death. Most others are consumed by bigger fish, and only a tiny fraction makes it to the next year. Small fish are involved in a race against the clock. It is crucial to grow as fast as possible for two reasons. First, larger animals fit in the mouth of a smaller fraction of the range of predators that are around. Second, as the fish become larger, more types of food items fit into their own mouths. Since larger individuals have a larger range of food available, they can grow faster. The resulting positive feedback can cause a conspicuous split between so-called *jumpers* that have become big enough to enter the positive feedback, and the *stunters* that do not make it to this point.[36] Their growth becomes stunted, and they are most likely to end up as food for bigger fish, including in some cases their brothers and sisters that became jumpers. The rapid growth of young fish and the effect of size on trophic interactions make the concept of *niche* of little use to describe the position of fish species in the ecosystem. Therefore, detailed models that focus explicitly on the dynamic feedback between individual size and foodweb interactions are often used in research on fish dynamics. Such models show that the size-dependent interactions in fish communities can cause a range of complex dynamical patterns, including alternative attractors[37] and complex multiyear cycles.[38]

Seasons are quite a different thing from the perspective of plankton. In episodes of rapid growth, phytoplankton can have a generation time of merely one day. This implies that for phytoplankton populations, the time between two winters is actually comparable to the period between two glacial cycles seen from the viewpoint of trees. As a result, complete succession dynamics play out over a growing season.[24] In spring, the rise of temperature and increased light availability trigger a spring peak of diatoms and other algal groups that take advantage of the nutrients that have become available from decomposition of dead organic material over the winter. The high algal abundance implies a wealth of high-quality food for zooplankton. Overwintering individuals start to grow and reproduce rapidly, and winter eggs (resting stages) also emerge to boost the population development further. Some weeks later, waterfleas have often become abundant enough to deplete their algal food to very low levels. The resulting period of low turbidity is known as the *spring clear water phase*. The clear water phase comes to an end as the waterflea populations collapse because of food shortage, and phytoplankton populations can once again become abundant. In the laboratory, this cycle of rise and fall of zooplankton and phytoplankton can go on forever. In fact, it is one of the classic examples of a predator–prey cycle (section 3.1). However, in the field, there are complicating factors. Usually, the young-of-the-year fishes have just grown to the size at which zooplankton is their preferred food by the time that waterflea populations would be ready to profit from the next phytoplankton peak and bounce back from their starvation collapse. Mostly, the young fish are then able to suppress waterfleas throughout the summer. Whether this happens depends critically on the amount of fish predation.[39] This critical switch point in the seasonal cycle can be seen as a *basin boundary collision* (section 3.3). The starvation crash that is part of the predator–prey cycle pushes the system over the boundary of the basin of attraction of a stable state in which waterfleas are overexploited by fish.[40] Subsequently, a succession of phytoplankton species begins that moves the community toward increasingly shade tolerant types. Usually, those are slow-growing species of cyanobacteria that form colonies. Between these "trees," there are small, fast-growing species that are grazed upon by zooplankton that is small enough to be an uninteresting prey for most fish. At the end of the summer when most young fish have died, waterfleas that are the large powerful grazers in this system can escape top-down control once more and graze away phytoplankton enough to create a fall clear water phase. Finally, winter comes, and productivity is reduced so much that only small populations of zooplankton and phytoplankton remain present.

Typical patterns that can be observed in seasonal variation in the total biomass of phytoplankton and zooplankton can be reproduced remarkably well by a very simple two-equation model of the interactive dynamics of these groups (figure 7.9). (A brief explanation of the model equations and the way in which the dynamics arise can be found in the appendix, section A.10.) The only requirement is that we let the parameters reflecting the effects of fish predation, light, and temperature vary in a sinusoidal way over the year (see the appendix, section A.10). The resulting seasonal patterns are mostly phase locked (see section 3.2) to produce a regular annual pattern but can also be chaotic (figure 7.9, bottom panel). Such patterns are indeed quite recognizable in practice, and depending on the nutrient level and fish density, all patterns illustrated in figure 7.9 can be predicted and found. However, these are patterns in which all species in a functional group are lumped. If we look at the species level, what we see is chaotic, and predictability becomes very low. This should be no surprise. Not only do many models of multispecies plankton interactions have chaotic attractors,[41,42] but also experiments with plankton in microcosms (figure 7.10) show irregular fluctuations that persist for years under constant external conditions.[43,44]

7.3 Other Alternative Stable States

So far, we have highlighted details of what happens in the clear and the turbid states. Those are states characterized by dominance by different primary producers (autotrophs): submerged vegetation for the clear state and phytoplankton for the turbid state. However, there are more classes of primary producers that may dominate lakes in a way that appears to have the character of a self-stabilizing condition in some lakes. Also, in deep lakes, a positive feedback between sediment nutrient recycling and phytoplankton productivity may cause alternative stable states. To complete our short tour of lake ecology, we will look into some of those alternative states now.

The Clear and the Turbid State Subdivided

Although there are many species of submerged macrophytes, one particular group differs rather fundamentally from the rest. Charophytes (stoneworts) are distinct not only taxonomically (they are algae rather

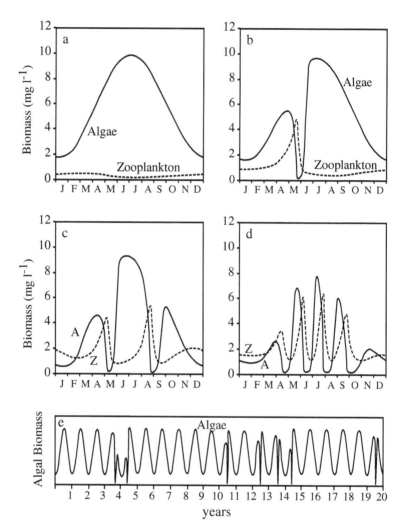

FIGURE 7.9. Different types of seasonal dynamics that can be observed in lake plankton generated by a periodically forced zooplankton–algae model. Panels (a) to (d) represent situations with increasingly low fish densities. Panel (e) shows chaotic dynamics for a fish density between that used in panels (a) and (b). (From reference 40.)

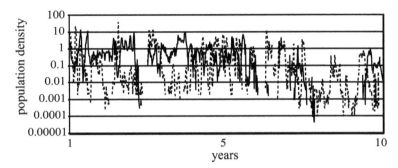

FIGURE 7.10. A plankton community kept in microcosms under constant conditions for ten years kept fluctuating chaotically, as exemplified by time-series of two of the species. (From reference 44.)

than "higher plants") but also ecologically. In some lakes, distinct shifts from pondweeds to an almost complete monoculture of charophytes have been observed.[14,45] This suggests that coexistence of these two groups may be unstable, and their competition might result in alternative stable states of dominance by either of the groups. (See the appendix, section A.4, for a general model of such competition.) To test this idea, Dutch ecologists have looked more deeply into the competition between two specific representatives of these groups: *Chara aspera Deth. ex Willd.* and pondweed *Potamogeton pectinatus L.* Indeed, it appears that there is an interesting asymmetry in the competition. The tall, canopy-forming pondweed is simply on top of the short charophyte, and therefore, pondweed is the better competitior for light.[46] However, it appears that the stonewort is strongly depleting bicarbonate and can at the same time survive at lower bicarbonate concentrations.[47] Thus, there is a positive feedback in the development of the stonewort as it drives the system toward carbon limitation, suppressing the pondweed, and thereby creating better light conditions for itself.[5] A detailed individual-based competition model suggests that this may plausibly lead to alternative stable states in the competition between these two submerged macrophytes.[4]

Just as it is too simple to consider submerged macrophytes as a single group, phytoplankton dominance may also take on different faces. Different species can rise to dominance, and despite intensive research and growing insight into the factors that drive succession, we are still far from able to predict which species will dominate when. Perhaps the best-studied group are blue-green algae. These are really bacteria (cyanobacteria) rather than algae, but as planktonic autotrophs, they occupy the same niche. Species in this group differ widely, and it is impossible to treat them as a single ecological entity. However, some groups have been studied more than others, as they are often a nuisance. In shallow lakes, filamentous cyanobacteria of the *Oscillatoria* group can be dominant all year round. These cyanobacteria are rather shade tolerant, which explains why they enter especially when the water has become sufficiently turbid.[48] Interestingly, they can also intensify the shady conditions once they are present. This is because with the same amount of phosphorus, they can build biomass that causes relatively high light attenuation.[49] As a result, they can stabilize their dominance once they are there, creating an alternative stable state in the phytoplankton community.[50]

In conclusion, the image of just two contrasting states is too simple. Different groups of primary producers may dominate shallow lakes, and such states dominated by a particular group may often represent alternative stable states. Shallow lakes may be dominated by charophytes, submerged angiosperms, green algae, or cyanobacteria (figure 7.11). Clearly, the shifts between two different types of submerged macrophytes or between two communities of phytoplankton are less spectacular than the shift between submerged macrophytes and a turbid phytoplankton-dominated state. Also, many more shifts in dominance may in reality exist. Therefore, the change of the biological communities along a gradient of eutrophication may really be seen as a continuum in which gradual species replacements are interrupted at critical points by more dramatic shifts to a contrasting community state (Figure 7.12), as is also shown in abstract models of complex communities of competing species.[42]

Free-Floating Plant Dominance in Ponds and Tropical Lakes

Although a turbid state with high phytoplankton biomass is the typical eutrophication problem in temperate lakes, invasion by mats of free-floating plants is among the most important threat to the functioning and biodiversity of freshwater ecosystems ranging from temperate ponds and ditches[51] to tropical lakes.[52] Dark, anoxic conditions

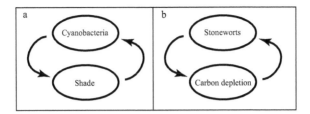

FIGURE 7.11. (a) Within the turbid state of shallow lakes, filamentous cyanobacteria may represent another stable state, as they can create shady conditions where they outcompete other phytoplankton. (b) Similarly, within the clear vegetated state, dominance by stoneworts can represent an alternative stable vegetation state, as they deplete bicarbonate to levels where it is difficult for their competitors to grow.

under a thick floating plant cover leave little opportunity for animal or plant life. Also, negative impacts on fisheries and navigation in tropical lakes can be dramatic.[52] Not surprisingly, resolving floating plant problems has a high priority in many warm regions of the world and is also a focus of quality management of many smaller water bodies in temperate regions. For example, in the extensive system of ditches and canals in The Netherlands, duckweeds are considered the main problem associated with eutrophication.[51]

It has been demonstrated that a shift to floating plant dominance in shallow water ecosystems is a critical transition.[53] Just like the shift to a turbid state in temperate lakes, the shift to floating plant dominance is difficult to reverse and can happen when a critical threshold level of nutrients is passed. The explanation is in the way the floating plants compete with submerged plants. Floating plants have primacy in competition for light but need high nutrient concentrations.[54] In contrast, rooted submerged macrophytes are susceptible to shading but are less dependent on nutrients in the water column, as they may take up a large part of their nutrients from the sediment.[55] Still-submerged plants can also use their shoots effectively for nutrient uptake from the water column[56] and by various mechanisms reduce nitrogen concentrations in the water column to below detection levels.[57] This interaction results in two alternative stable states: a floating plant–dominated state in which invasion by submerged plants is prevented by shading, and a situation dominated by submerged plants in which invasion by free-floating plants is prevented by reduced nutri-

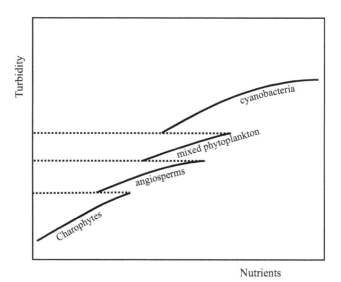

FIGURE 7.12. Schematic representation of how the classic graphical model of alternative stable states in shallow lakes can be expanded for temperate shallow lakes to cover four alternative equilibria. For tropical lakes, dominance by free-floating plants may be yet another alternative stable state (see text).

ent availability.[53] This bistability makes reversal of floating plant invasion of a lake often difficult. On the other hand, it implies that if nutrient levels have been sufficiently reduced, a one-time removal of floating plants might tip the balance to an alternative stable state dominated by submerged plants. The evidence for the idea that floating plant dominance can be an alternative attractor is based (as in the case of the turbidity shift) on a combination of field patterns, mechanistic models (see appendix, section A.13), and experimental results (figure 7.13).

ANOXIC PRODUCTIVITY SHIFT IN DEEP LAKES

So far, most of the lake dynamics that I have shown come from shallow lakes, my own favorite ecosystem for many years. However, before moving from lakes to a review of critical transitions in the Earth system, I should mention that deep lakes can be bistable too.[58] The mechanism

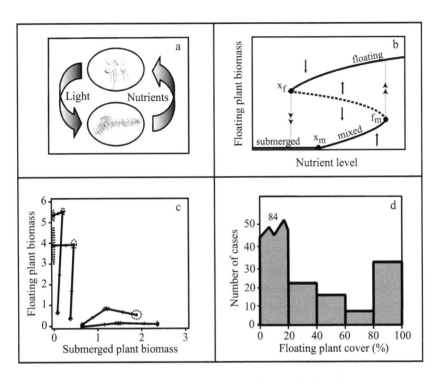

FIGURE 7.13. Floating plant dominance and submerged plant dominance are thought to represent alternative stable states in ponds, ditches, and tropical lakes. This has been inferred from various lines of evidence: models that represent the asymmetric competition for nutrients and light (a) predict alternative attractors for a large range of parameter settings (b); experiments starting from different initial states converge to dominance by either of the groups (c); and data from Dutch ditches show a bimodal distribution of abundance of floating plants. (Redrawn from reference 53.)

here is that as the lake becomes enriched, the phytoplankton productivity increases, leading to a larger load of organic material settling to the sediment. Because in such lakes, the deep, dark water layers are not mixed with the surface water in summer, the deepwater oxygen can become depleted, and this happens more rapidly if more organic material needs to be decomposed, consuming the oxygen. As a result, lakes with more phytoplankton have higher risks of anoxia. Since sediments start to release their stored phosphorus under anoxic conditions, this implies that more phosphorus becomes available for the phytoplankton, causing even higher productivity, leading to more anoxia, and so on.

7.4 Synthesis

Lake ecosystems provide well-documented examples of critical transitions, but they also illustrate why the idea of two alternative stable states is really a caricature of reality. First, more than two alternative states appear to exist, some more different than others. Therefore, as conditions change, the system will show gradual replacement of species, occasionally interrupted at critical points by smaller or larger jumps in the community. Second, *stable* is a difficult concept in practice. Lakes that are frequently hit by extreme events such as desiccation or solid freezing may be relatively far from equilibrium most of the time. Also, lakes may under some conditions display multiyear cycles back and forth between the clear and the turbid state driven by intrinsic mechanisms. Last, in temperate regions, the clear and the turbid state consist of seasonal cycles that are relatively predictable on the level of functional groups, although there are critical switch points in the dynamics of fish and plankton, and multiyear cycles and complex dynamics can arise in fish and plankton communities. Viewed on the species level, plankton dynamics are mostly chaotic. Thus, although the image of the clear vegetated state and the turbid state as alternative attractors is a useful characterization for understanding the large-scale behavior of shallow lakes, much is going on underneath this rough level of classification.

CHAPTER 8
Climate

Much like a lake, our planet is an almost closed system. Radiation passes in and out, but for the rest, the system is almost self-contained. Although these dry facts were known before, a true paradigm shift was triggered when the first astronauts saw and photographed the Earth from space. They were overwhelmed by the image of what seemed a "living thing." Just as the image of a lake invoked the science of ecology, the pictures of the Earth from space inspired scientists to develop an integrated view of our planet and the mechanisms maintaining its dynamics. A first influential idea along those lines was James Lovelock's *Gaia hypothesis*, proposing that life as a whole fosters itself by helping to create an environment on the Earth suitable for its continuity.[1] He hypothesized that the living matter of the planet functioned like a single organism and named this self-regulating living system after the Greek goddess Gaia. Although the Gaia hypothesis has been associated with quasi-mystical thinking about the planet Earth, Lovelock's ideas clearly stimulated thinking about feedbacks in an increasingly important field of science, focused more and more on predicting how the Earth system may respond to the ongoing changes in atmospheric composition and land use imposed by humans.

Most climate studies up to now predict gradual change. However, after reading this chapter, you should be convinced that gradual trends will inevitably be interrupted by occasional drastic shifts to a contrasting climate regime. While the original Gaia idea stressed the stabilizing feedback of life on the abiotic conditions, it has become apparent over the years that the Earth system is characterized by stabilizing (negative) but also by some strong positive feedbacks. Like a lake, the Earth is a complex nonlinear system, and reconstructions of past changes suggest that just like a lake, its dynamics are characterized by regime shifts, oscillations, and chaotic fluctuations (the hallmarks of such nonlinearity) on all timescales.

When Forbes wrote "The Lake as a Microcosm" in 1887, he hinted that a lake system may somehow reflect what happens in the world as a whole.[2] He cannot have imagined how remarkable some of the parallels between the dynamics of lakes and those of the Earth system would turn out more than a century later. Despite the huge difference in scale, many similarities in the way the interplay of internal feedbacks and external drivers causes regime shifts cycles and chaos are striking. This is why I jump from lakes to the Earth system, rather than going through the intermediate-size examples such as terrestrial ecosystems and oceans first.

I start in *deep time*, where only massive events have left traces that are profound enough to allow reconstruction today, and proceed through the Pleistocene glacial cycles and more recent spectacular events like the Younger Dryas, to end with a look at present-day climatic phenomena such as El Niño events and prolonged droughts like the infamous Dust Bowl that hit the Great Plains of the United States in the 1930s.

Importantly, the explanations of those climate shifts are much more speculative than the theories about the dynamics of lakes. There are several reasons for this. First, most of the past climate events we try to study happened only once or a few times, whereas critical transitions and cycles in lake ecosystems can be studied over and over. Second, while we can study lake dynamics in great detail, our view of past climate dynamics is a based on imperfect reconstructions that have to make use of indirect evidence from sources such as tree rings, sediments, rocks, and ice-cores. Third, our best understanding of what drives the dynamics of lake ecosystems comes from well-designed experiments on scales varying from enclosures to entire lakes. Clearly, experimental manipulation of the climate system on meaningful scales is out of the question. As a consequence, the history of scientific inquiry into ancient climate events often reads like a detective story, and rarely do we get close to certainty when it comes to the question "Who did it?" What I present here is the state of the art, but even as I write this, new articles keep appearing that challenge the previous views.

8.1 Deep Time Climate Shifts

The stretch of the history of our planet far before humans were present is often referred to as *deep time*. Like deep space, it is a concept that is difficult to conceive, as it is so far beyond what we can imagine intuitively. Yet science can handle those different timescales quite well. For instance, we can estimate aspects such as the age of our planet using rigorous scientific approaches. Also, we can reconstruct the dynamics of climate and other characteristics of the Earth from sediments and other remains. As you will see, the magnitude of some of the ancient shifts in the Earth system suggested by those reconstructions is truly impressive.

The Great Oxidation

The largest chemical transition in the history of the Earth was the so-called *Great Oxidation*. About 2.4 billion years ago, the atmosphere shifted suddenly and irreversibly from a state with about 10^{-5} times the present atmospheric level of oxygen to one with 1 to 10% of the current level. The simplest explanation of this more that 1,000-fold increase would be the evolution of cyanobacteria that could do the trick of oxygenic photosynthesis. However, this happened 300 million years earlier. So what kept oxygen levels low for so long? Another puzzling aspect is the abrupt nature of the shift. Proxies indicate that at either side of the Great Oxidation, oxygen concentrations were relatively stable for $10^8 - 10^9$ years. So why such a sudden jump between widely different levels?

Part of the explanation may be that the first oxygen produced was simply used up for oxidation of iron and other easily oxidized compounds of the Earth's crust.[3] However, a recent analysis suggests that this cannot be the whole story.[4] Instead, it shows that the early atmosphere should be expected to have had two alternative stable states, one low in oxygen and the other with a high oxygen concentration. Here, I summarize this view portraying the Great Oxidation as a critical transition.

In shaping the ancient atmosphere, as today, organisms played a pivotal role. Photosynthesis worked just the same way, using energy from light to make carbohydrates from water and carbon dioxide. The principal fate of those carbohydrates in the current world is *aerobic respiration*. This is just the reverse process, releasing energy and breaking down the carbohydrates into water and carbon dioxide again. However, this did not work in the ancient atmosphere, as this process is inhibited if there is less than 1% of the current atmospheric oxygen in the air (the *Pasteur point*). In that situation, the carbohydrates are broken down into a mix of methane (CH_4) and CO_2. The net result of the photosynthesis and this *anoxic respiration* before the Great Oxidation must therefore have been a flux of methane and oxygen into the atmosphere. This mix of methane and oxygen is far from thermodynamic equilibrium. The dominant process in restoring equilibrium is atmospheric methane oxidation, transforming the methane and oxygen back to CO_2 and water again. This takes place as a series of reactions, some of which are photochemically mediated.

Such methane oxidation must have consumed most of the early oxygen delaying the Great Oxidation for 300 million years. So what eventually made this process change, triggering the Great Oxidation then? The key to the explanation is the fact that once oxygen has increased to a certain level, an ozone layer forms, shielding the troposphere from ultraviolet radiation. This dramatically decreases the rate of photolysis of water vapor in the troposphere, a procees that produces hydroxyl radicals that are essential for the methane oxidation. Combining the essential chemical reactions in a dynamical model suggests a *catastrophe fold* in the equilibrium curve of atmospheric oxygen (figure 8.1). Either a slow decrease of reducing matter or a gradual increase of net photosynthetic oxygen input would eventually cause the system to reach a tipping point at which a jump to 1,000-fold higher oxygen concentrations is predicted (note the logarithmic scale). A closer look at the available data suggests that the Great Oxidation was probably triggered by a decrease in reductant (like iron), moving the system from a situation with only one stable state to the bistable region. The actual shift may well have been triggered by a

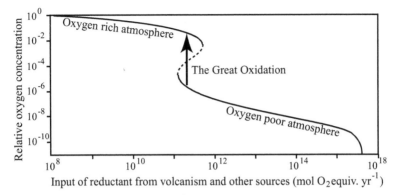

FIGURE 8.1. Stable states of the atmospheric oxygen concentration as a function of the input of reductants. The Great Oxidation, at which the concentration jumped more than 1,000-fold, probably happened as the result of a minor perturbation of the carbon cycle, after a decrease in the input of reductants such as ferrous iron had brought the system into a situation with two alternative stable states. (From reference 4.)

small carbon cycle perturbation such as a temporary increase in burial of organic matter. The transition would then take 50 to 150,000 years—a swift event compared to the 300 million years of waiting time after the onset of oxygenic photosynthesis.

The Great Oxidation has been a prerequisite for the later spectacular development of larger life-forms (section 9.2). However, it probably had other dramatic side effects first. In addition to creating an oxygen-rich atmosphere, it implied a large decline in atmospheric methane concentration. Since methane is a powerful greenhouse gas, this implied a dramatic drop in temperature. The associated temperature drop is estimated to be 4–9°C, and this may well have been sufficient to trigger "snowball Earth," the most extreme glaciation in the history of the globe, as explained in the next section.

THE SNOWBALL EARTH GLACIATIONS

In the 1960s, a remarkable hint on the climate in the far past was found. It turned out that glacial deposits dating from the end of the Neoproterozoic eon (1000 to 543 million years ago), could be found on virtually all continents, including in tropical areas. This led to the suggestion that once the Earth had been entirely covered by ice.[5] It was difficult to really prove that in those days, as there was also the possibility that various continents were glaciated at different times at moments when continental drift had simply brought them close to the poles.

At the same time, however, the first mathematical models of the Earth's radiative balance and the mechanisms governing glaciations were developed. Surprisingly, those simple models also pointed to the possibility of a complete glaciation of the planet. The mechanism is easy to understand. The Earth's temperature is fundamentally controlled by the way that solar radiation is processed. About one third of the incoming solar radiation is reflected back to space by clouds and by the Earth's surface, and the rest is absorbed. The absorption increases the average temperature and causes the Earth's surface to emit infrared radiation, balancing the energy that has been absorbed (depending among other things on the amount of greenhouse gases). Clearly, if more of the solar radiation is reflected in the first place, less radiation is absorbed, and the Earth's temperature becomes lower. The *albedo* is a measure of reflectivity; snow has a high albedo (about 0.8), and seawater a low albedo (about 0.1). Land surfaces have intermediate albedo values. Straightforward reasoning thus tells us that a world with snow and ice reflects more radiation into space and therefore will become colder. Since a colder climate in turn promotes snow and ice, this represents a positive feedback in the Earth system (figure 8.2).

It is intuitively straightforward that this might in principle lead to a runaway process in which the world gets colder and colder until everything is covered by ice. However, would this really be a plausible scenario?

The Russian pioneer of climate science Mikhail Budyko was the first to design a simple energy-balance model showing that the ice–albedo feedback might indeed lead to a global glaciation.[6] He estimated that the critical threshold for that would be reached if ice covered about half the Earth's surface area (figure 8.3). Beyond that point, the feedback would cause a further shift toward a completely frozen planet. On such a world, only the little bit of heat escaping from the Earth's interior would prevent the oceans from freezing to the bottom, and a massive layer of sea ice would form from the poles to the equator.

Still, few in those days believed that the Earth had really once been transformed into a "snowball." There were two reasons for this. First,

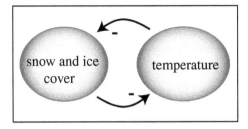

FIGURE 8.2. The ice–albedo feedback. Low temperatures lead to more snow and ice cover, but the reflection of radiation by snow and ice back to space leads to further cooling.

it seemed that such a catastrophe would have extinguished all life, whereas evidence exists of extant life-forms long before the times corresponding to the worldwide glaciation marks. Second, it was unclear how the Earth would ever be able to return from such a frozen state. The first problem disappeared with the discovery of organisms living in deep-sea hot-water vents and in the extremely cold valleys of Antarctica. Some of these organisms indeed seemed to have existed in the Neoproterozoic eon before the supposed glaciations. The second problem can be resolved if one accounts for volcanism, a process that may have slowly undermined the stability of the frozen state.[7] Over very long timescales, the amount of carbon dioxide in the ocean–atmosphere system results largely from supply by volcanoes on one side and removal by chemical weathering reactions with silicate rocks on the other side. Now, if the Earth were so cold that there could be no liquid water on the continents, weathering reactions would stop. However, volcanoes would keep pumping CO_2 into the atmosphere, allowing carbon dioxide to build up slowly to higher and higher levels. Eventually, the carbon-dioxide-induced greenhouse effect would become so large that it would compensate the ice–albedo effect and the ice would start to melt.

Given that solar power in those days was about 6% lower than today, it was estimated that roughly hundreds of times the present concentration of atmospheric CO_2 would have been required to overcome the albedo of a snowball Earth.[8] This incredibly high value has two major implications. First, extrapolating from the current rates of volcanic carbon dioxide emissions, a snowball Earth state would have

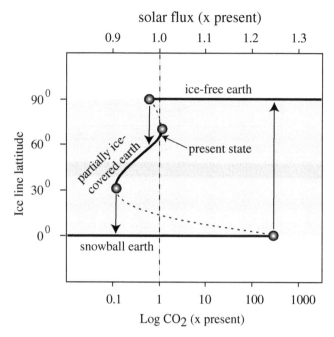

FIGURE 8.3. Stable (solid) and unstable (dashed) extents of ice cover as a function of CO_2 levels or solar radiation predicted from a simple energy balance model of the Earth. The dots represent bifurcation points at which the system jumps to an alternative equilibrium. The vertical arrows indicate the cycle of transitions that may have occurred several times in deep time. Note that this (overly simple) model suggests that the present state of the Earth could be close to a bifurcation that gives rise to an ice-free Earth. (Based on a Budyko-Sellers-type energy balance model, modified from *http://www.snowballEarth.org*, accessed January 2007.)

lasted for millions to tens of million of years before becoming unstable. Second, once the runaway process of melting was triggered, the Earth would become tremendously hot because of the greenhouse effect of such high CO_2 levels. The warming is further promoted by evaporation adding water vapor to further boost the greenhouse effect and produce ocean temperatures of almost 50°C.[9]

Although it remains difficult to pin down what happened 600 to 700 million years ago, later work has provided various lines of geological evidence that the Earth has indeed been frozen to the equator and that the scenario sketched here might well represent the way it happened. Also, it appears that not one but several such glaciation cycles happened, "freezing and frying" the Earth in sequence. Nonetheless, the puzzle is not yet completely solved. In addition to the question of how such massive glaciations ended, there is the question of how they started in the first place. The fact that solar radiation was less in those days helps a bit, but most studies agree that very low atmospheric CO_2 levels would be needed to make the Earth cold enough to trigger the full runaway glaciation. A central question therefore is what process might have caused CO_2 concentrations to have dropped so deeply. There are two major candidates. First, some time before the first snowball Earth glaciation, the supercontinent Rodinia broke up into smaller parts. This may have led to an increased CO_2 binding for several reasons.[10] A broken-up continent implies more precipitation and runoff on land compared to the climate on the huge supercontinent, because of the increase in sources of moisture along continental borders. This allows more silicate weathering, which consumes CO_2. In addition, the breakup of the supercontinent into smaller plates is accompanied by the eruption of large basaltic provinces, resulting in an increase of weatherability of the continental surface. The runaway glaciation would be more likely to result from these processes if the new continents became concentrated at low latitudes where they could keep consuming CO_2 longer than if they would be close to the poles, where they would start to be covered by ice caps earlier on, halting their CO_2 consumption. As mentioned earlier, another mechanism that may have played a role in triggering the first snowball Earth runaway glaciation is the evolution and radiation of carbon-fixing cyanobacteria,[11] which resulted after some time lag in the oxidation of the atmosphere, causing a sharp drop in concentration of the powerful greenhouse gas methane.[4] In this view, an evolutionary accident that created photosynthesis eventually led to one of the world's worst climate disasters, nearly bringing an end to life altogether; a dramatic example of the intimate interaction between the evolution of life and climate.

THE PALEOCENE-EOCENE THERMAL MAXIMUM

After the last snowball state ended about 600 million years ago, the Earth stayed in a largely ice-free greenhouse state until only 35 million years ago, when the current "icehouse" condition started. This very long warm period, which allowed the evolution of ancient plants, dinosaurs, and most current life-forms, has not passed smoothly all along. For instance, about 55 million years ago, a brief but extreme warming event occurred, probably triggering the remarkably simultaneous evolution and radiation of important orders of mammals such as the deer, the horses, and the primates around that time.[12] The obvious sharp changes around this period had already led scientists to classify it as the boundary between two distinct geological periods: the Paleocene and the Eocene. The abrupt warming event that later turned out to have marked this transition is therefore called the Paleocene–Eocene Thermal Maximum (PETM). Various explanations of the abruptness of this event have been suggested. One possibility, suggested by analysis of isotopic signatures of global deposits, is that the PETM is related to the release of large quantities of methane from marine sediments on ocean margins where it was buried as methane hydrate.[12] Again, this could be a runaway process due to a positive feedback on the Earth's temperature. Methane is a greenhouse gas driving up global temperatures. This promotes the release of more methane, as the methane hydrates become unstable at higher temperature. Only when the methane reservoirs became exhausted would this process slow down and stop. The PETM thus ranks in the growing collection of examples of past rapid climate change that may be related to situations in which a runaway process leads to a sharp climatic change once a critical threshold is passed. It has been suggested that astronomic pacing was involved in triggering this extreme warming event as well as a possible second event 2 million years later, as both events coincided with maxima in eccentricity cycles of the Earth's orbit.[13]

FROM GREENHOUSE TO THE CURRENT ICEHOUSE

Eventually, the long, warm greenhouse period that followed the snowball Earth glaciations ended and gave way to the icehouse climate in which we live now, a period in which ice caps started to appear once again, albeit in a more modest form than during the massive global glaciations in deep time. The icehouse climate is characterized by an oscillation between glacial and interglacial periods and started about 34 million years ago. Before turning to the mechanisms that drive the cycles of glaciation, it is worth noting that the transformation from the greenhouse to the icehouse world (the Eocene–Oligocene transition) was another sharp transformation in the Earth's climate. The underlying cause must have been a decrease in CO_2 levels. This may have been related to the growth of the Himalayas, resulting in enhanced weathering of the rocks unearthed, combined with increased biological productivity fixing CO_2 through photosynthesis.[14] Interestingly, it appears that the shift to the permanent oscillating icehouse climate seems to have been preceded by several small glaciations and one major transient glaciation.[14] This confirms the image predicted by simple models (figure 8.3) that the icehouse and the greenhouse state may be alternative attractors. Around the transition from Eocene to Oligocene, the Earth was apparently close to the bifurcation points of this small hysteresis, and the resulting small basins of attraction may have allowed perturbations such as volcanic eruptions and variation in the Earth's orbit to invoke jumps back and forth between the states.[15] Eventually, the CO_2 levels would have fallen enough to preclude the Earth from being tipped back into the greenhouse state, implying permanence of the ice caps that have covered the Earth since then. Note that even if the range of CO_2 levels for which alternative attractors exist were very small, ghost effects (see section 5.1) may have produced prolonged transients upon perturbations that might explain such transient precursor glaciations.

8.2 Glaciation Cycles

The spectacular deep time events we have examined up to now can be reconstructed only coarsely. However, as we get closer in time to the present, the resolution of our image of past climate dynamics becomes increasingly detailed. The cyclic waxing and waning of ice caps over the past million years (the Pleistocene) is probably the best-known example of strong climatic change, and this kind of change is not far behind us. Archeological remains such as cave paintings vividly remind us that even our relatively recent ancestors, modern humans in all physical aspects who produced wonderful stylized paintings, had to deal with a situation in which much of Europe and North America was covered by massive glaciers. Here I will first look at the pattern of

Pleistocene glaciation cycles and then discuss the mechanisms that appear to have driven the rhythm of the climate.

Ice sheet variations over the last, say, two million years can be read from isotopic signatures of ocean sediments, and climate dynamics over the last half million years can be reconstructed from ice-cores drilled from the ice masses up to 3 km thick that still cover the poles. The well-known images of those climatic reconstructions show repeated cycles of slow cooling interrupted by relatively sharp episodes of rapid melting (figure 8.4; note that time goes from right to left in most paleo-figures).

THE MILANKOVITCH CYCLES AS DRIVERS OF GLACIAL CYCLES

The explanation of glacial cycles is simple at first glance. The warm episodes in the glacial cycle pattern correspond roughly to maxima in insolation (figure 8.4). This variation in insolation is due to variations in the orbit of the Earth caused by astronomical factors such as the influence of other planets. Although the rhythm of waxing and waning of ice sheets does not simply track the fluctuations in insolation, the Vostok data and other sources do suggest a relationship between the glacial periods and the changes in the Earth's orbit around the sun.[17]

Although the variability of the Earth's motion is quite complicated, it consists largely of quasi-periodic changes in three major aspects: the eccentricity of the orbit, the axial tilt, and precession (wobble). Each of these aspects varies with a characteristic frequency, and taken together those cyclic variations, known as the Milankovitch cycles, translate into alterations in the radiation reaching the Earth's surface and in the contrast between seasons.

The orbital shape ranges between more and less elliptical in a cycle of about 100,000 years. This modulates the amount of radiation received at the Earth's surface over a year. In the current condition, a difference of only about 3% occurs between farthest point and the closest point, causing us to receive a mere 6% more solar energy in January than in July. However, when the Earth's orbit is most elliptical, the seasonal difference in solar energy is 20 to 30%.

Our seasons are due to the tilt of the Earth's axis relative to its plane of orbit around the sun. Oscillations in the degree of this incli-

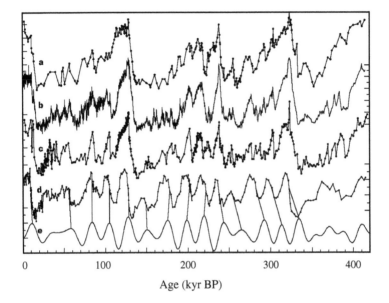

FIGURE 8.4. Time-series of climate and atmospheric compositions reconstructed from the Vostok ice core plotted together with variation insolation resulting from variation in the Earth's orbit: (a) CO_2, (b) atmospheric temperature, (c) methane, (d) isotopic signature of ice indicating local temperature when it was formed, and (e) summer insolation at 65°N. (From reference 16.)

nation occur with a periodicity of 41,000 years. Today's tilt is approximately in the middle of the possible range. Less tilt implies a smaller amplitude of seasonality but also a larger difference between the average radiation received by the polar regions as opposed to the equator. This might promote the growth of ice sheets, as warmer winters would result in more snowfall and cooler summers would imply less melting.

The third Milankovitch cycle corresponds to a wobble in the Earth's spin axis of about 23,000 years. At one extreme of this wobble, the Northern Hemisphere experiences winter when the Earth is farthest from the sun and summer when the Earth is closest to the sun. This results in greater seasonal contrasts. At present, the Earth is closest to the sun in Northern Hemisphere winter, thus ameliorating the seasonal contrast.

AMPLIFYING FEEDBACKS

In spite of the apparent and intuitively straightforward link between variation in radiation and ice ages, the response of the Earth to the Milankovitch cycles appears rather difficult to explain after a closer look.[18] For one thing, the variation in incoming radiation resulting from the cycles is really not so large. Even when all of the orbital parameters favor glaciation, the increase in winter snowfall and decrease in summer melt seems not enough to grow large ice sheets. This happens only because of amplification of the forcing by positive feedbacks.

The most obvious amplifying feedback in this case is that snow and ice absorb much less radiation than ground and vegetation. Thus, ice masses reflect more radiation back into space, cooling the climate and allowing glaciers to expand further. As discussed earlier, this can in principle cause a runaway process leading to complete coverage of the Earth by snow and ice, something that may have happened on a few occasions in deep time. The necessity of this feedback for amplifying the radiative forcing signal also implies that the position of the continental plates on the surface of the Earth matters. On very long timescales, continental drift changes this distribution. When land masses are concentrated near the polar regions, as they are today, snow and ice can accumulate better, and the Earth is more prone to glaciations. In contrast, glaciations may occur less readily at times in the history of the Earth with less land in the polar regions.

Another important class of amplifying feedbacks can be seen from the ice-core reconstructions of atmospheric change (figure 8.5). During warmer periods, concentrations of CO_2 and CH_4 (methane) were higher. Most likely, those atmospheric changes result from the variation in temperature. For instance, high-resolution data from the termination of the last glacial cycle and, on a faster timescale, data from the recent cold period known as the little ice age show that the gas concentrations lag behind the temperature increase a bit. However, since such greenhouse gases promote higher Earth temperatures too, the fact that the gases themselves are temperature driven implies a positive feedback.[19]

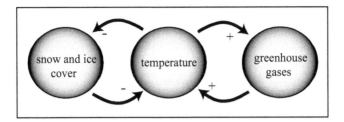

Figure 8.5. Two of the feedbacks that amplify the effect of subtle variations in insolation and other factors to produce marked climatic changes such as glacial cycles. Such positive feedbacks also amplify the effect of emissions of greenhouse gases and changes in land use to produce a larger global warming than would be expected from greenhouse gas concentrations alone.

Why Always Slow Cooling but Rapid Warming?

A remarkable and important feature of the ice sheet dynamics that makes them look different from the Milankovitch cycles that drive them is their asymmetry. Ice sheets grow very slowly, which can be understood given the time it takes to build up an ice layer a few kilometers thick from snowfall. In contrast, terminations of glacial periods have invariably been relatively fast over the last 400,000 years. Looking at the glacial cycles, it seems as if the ice sheets build up until a critical point, at which they become *unstable*, and events such as a peak in radiation due to the orbital situation can trigger a self-catalyzing meltdown. The reflectivity feedback and the greenhouse gas feedback may help explain such rapid melting. However, it is believed that there are other mechanisms promoting rapid melting too. One theory that may explain why very thick ice caps may suddenly become unstable suggests that as ice sheets become thicker, the pressure at the contact layer with the bedrock increases. As a result, the pressure melting point can be exceeded, and the ice mass starts sliding on its liquid base downhill into the sea, where it melts relatively rapidly.[20]

Recently, glaciers have been observed to be melting, as well as sliding into the ocean more rapidly in several places. Clearly, this is a worry, as it may result in rapid sea-level rise. Several positive feedback mechanisms have been proposed to boost glacial melt in this context.

First, there is the important phenomenon that reflectivity of the ice decreases as it starts melting. This makes the glacier capture more heat and melt faster. A factor that may be influential for glaciers that reach the sea is retraction of the end of the glacier. This may allow seawater to enter between the glacier's former end-moraine and the ice, promoting faster melting. It also promotes retraction of the grounding line, reducing resistance and facilitating more rapid sliding into the sea.[21] Last, it is now thought that melting water trickling down the cracks in the ice may be lubricating the movement of melting glaciers[22] (figure 8.6).

Clearly, these and other positive feedbacks in the climate of the Earth are relevant not only for understanding how past glacial cycles have emerged, but also for assessing the effect of future global warming resulting from greenhouse gas emissions and altered land use. Just as minor variations in solar radiation captured by our planet have been amplified by positive feedback mechanisms, current human impacts on the climate may be amplified. One of the major challenges in predicting future warming is to find out how the numerous negative and positive feedback mechanisms may precisely moderate the dynamics.

Mechanisms that Alter the Rhythm

Although positive feedbacks may explain why marked climatic oscillations such as glacial cycles may result from apparently rather subtle forcing, they do not explain why the match between the Milankovitch cycles and the glaciation cycles is so messy. The timing of climatic extremes seems only loosely linked to the insolation rhythm (figure 8.7), and temperature dynamics appear to have various frequencies that appear to be absent or only very weak in the Milankovitch cycles.[23] One of the most puzzling aspects emerging from long-term reconstructions of the ice volume from ocean sediment isotopic signatures is that a marked 100,000-year periodicity suddenly came to dominate the pattern a million years ago, even though the orbital forcing pattern has not changed[20] (figure 8.7). The reason for this *Mid-Pleistocene Transition* (MPT) remains unresolved, but it may be due to long-term changes in overall CO_2 concentrations or to erosion of the *soft beds*

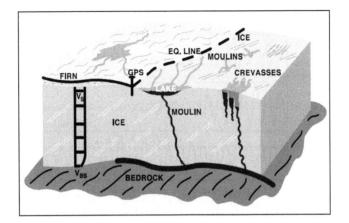

FIGURE 8.6. Melting water from the surface of large land–ice masses may reach the near-basal boundary layer, where it can "lubricate" the movement, promoting the glaciers to slide more rapidly into the sea. (From reference 22.)

underlying the ice sheets in the previous period.[20] In the absence of soft beds, the ice sheets do not spread out so thinly anymore. The supposedly thinner ice caps before the transition would respond relatively directly to the orbital forcing. This implies that they could follow the dominant forcing frequencies of about 20,000 and 40,000 years. In contrast, after the transition, the ice sheets supposedly could not spread out so thinly, because the lubricating weathered soft beds that had accumulated in the millions of years of warm climate before the current icehouse conditions had disappeared. The new state would be characterized by thicker, and therefore less responsive, ice sheets.

Although this sounds reasonable, how can a thickening of the ice caps result in such a sharp regime shift in fluctuation patterns, bringing out a low frequency that is not strong in the orbital forcing at all (figure 8.7)? Models suggest that this may have to do with an internal tendency for oscillation in the climate system, which is affected by the sluggishness of the response of thick ice caps.[24,25] The Earth system has some intrinsically slow variables, such as ice sheets and oceans, that through delayed feedback processes may produce intrinsic oscillations even in the absence of orbital forcing. Here is one example: When the hydrological cycle is most active, much precipitation is

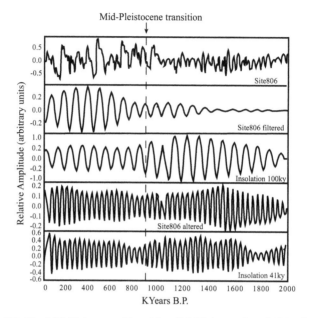

FIGURE 8.7. The Mid-Pleistocene Transition (MPT) is a regime shift in the Earth system about 1 million years ago at which the glaciation cycles changed in amplitude and became dominated by a 100-ky frequency. Top panel: Ice volume reconstructed from isotopic signature of ocean sediments. Second panel: These data filtered to highlight the 100-ky frequency. Third panel: Insolation forcing filtered the same way for comparison. Bottom two panels: The same, filtered to highlight the 41-ky cycle. (From reference 23.)

produced, and snow makes ice sheets grow. Eventually, however, this produces a colder Earth with less precipitation. There is a delay in the response, as ice sheets keep growing in spatial extent even if snow addition stops, because the thick ice caps spread out gradually. Nonetheless, after some time with little snow, the ice sheets start shrinking. When the ice sheets have become small enough, precipitation starts to increase, promoting ice sheet expansion again. Despite the growing ice sheets, the temperature remains relatively high for some time because of the inertia in ocean temperatures. This allows the hydrological cycle to remain active and causes snow accumulation to proceed for some time.

Combining such mechanisms that drive internal Earth system oscillations with the periodic orbital forcing, models can generate realistic-looking climatic dynamics that contain the 100,000-year frequencies as well as faster and relatively wild and irregular features characteristic of the reconstructed time-series of glacial fluctuations.[25]

Just as in lake plankton, the key to understanding the mix of regularity and chaoticity in the fluctuations is unraveling the effect of an external forcing function on an internally ticking "clock."

8.3 Abrupt Climate Change on Shorter Timescales

Zooming in further on recent timescales, we find abrupt climatic shifts that modern humans have faced. Civilization really evolved in a period of surprisingly constant climate, known as the Holocene, that started 10,000 years ago after the last glaciation. We will first take a look at a dramatic bump along the way from this glaciation to the long benign climatic period that followed. Subsequently, we will take a closer look at the more subtle ripples in the climate of the Holocene, such as the El Niño Southern Oscillation (ENSO) and sudden episodes of prolonged droughts.

DISRUPTIONS OF THE THERMO-HALINE CIRCULATION

Surprisingly abrupt events appear to have interrupted the last glacial period and the return to warmer times at the end of that glaciation. Perhaps the best-known example is a cold interval that interrupted the recovery from the last ice age (figure 8.8). Reconstructions from Greenland ice-cores suggest that this period, known as the Younger Dryas, was extraordinarily cold, dry, and windy in the Northern Hemisphere. While the cooling that started this period was quite irregular, the warming at the end happened through a single spectacular jump of more than 10 degrees Celsius. In fact, a similar heating jump ended the major glaciation about 3,000 years earlier, and as mentioned before, abrupt heating versus somewhat more gradual cooling are typical over all of recorded climatic history.

The Younger Dryas is only one example in a longer list of relatively brief but dramatic excursions of the climate system that have been reconstructed. Several of these events have been related to shifts in oceanic circulation patterns, especially the so-called *thermo-haline*

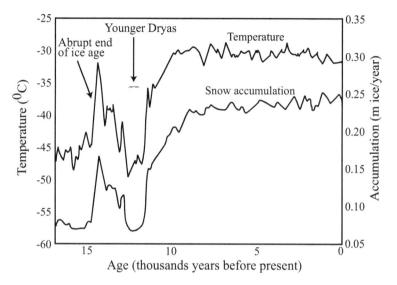

Figure 8.8. Climate dynamics reconstructed from central Greenland ice-cores reveals an abrupt warming that ended the last glaciation about 15,000 years ago, a cooling with sharp irregular lapses into the cold so-called Younger Dryas period, and subsequently another abrupt warming that ended this cold episode. (From reference 26.)

circulation. This circulation, also known as the "conveyor belt," transports heat from the tropics to the North Atlantic, and therefore allows the temperature in Northern Europe and North America to be considerably warmer than it would be without this "central heating system." The way it works (figure 8.9) is that cold water sinks in the North Atlantic as it becomes denser because of cooling. When it travels throughout the world's oceans toward the equator, it gradually warms, becomes less dense, and mixes to the surface. As it moves back along the surface of the ocean, it carries heat absorbed along the way. The reason that this system can cause abrupt climatic change is that it has alternative stable states. If a critical threshold is exceeded, the circulation may therefore shift relatively abruptly between an "on" or an "off" state. The positive feedback that is behind this bistability has to do with the salinity of the water. Put simply, inflow of freshwater into the North Atlantic makes it more difficult for the water in this region to sink because freshwater is lighter than saltier water. As this slows

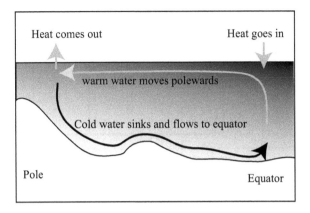

FIGURE 8.9. Basic way in which the thermo-haline circulation transports water from the equator to the North Atlantic. If freshwater enters in the high-latitude ocean, this reduces the density of the water and may prevent it from sinking, causing the current to switch off.

down the current, the freshwater is not transported away with the current, and the northern waters become increasingly fresh, until the point that the conveyor belt becomes blocked. Although later studies revealed many subtleties in this spatially complex system,[27] the basic image of how it operates was suggested already in the early 1960s[28] and still appears largely correct.

The Younger Dryas and various other sudden cooling events in the Northern Hemisphere seem to have resulted from a switch-off of the thermo-haline circulation triggered by floods of glacial meltwaters that freshened the Atlantic. Just like other systems with alternative attractors, the stability of the alternative states depends on various factors. In this case, temperature and salinity are the two major aspects.[29] The present-day climate corresponds to an active circulation (figure 8.10a). Only if freshwater input to the North Atlantic exceeds a certain threshold does the circulation switch off. If that happens, even if the freshwater input returns to the normal value afterward, the system will remain on the lower branch of the hysteresis curve and remain switched off. Only if freshwater is removed at a sufficient rate (for example, through evaporation) is another threshold passed at which the circulation switches on again. Note that the predicted hysteresis is

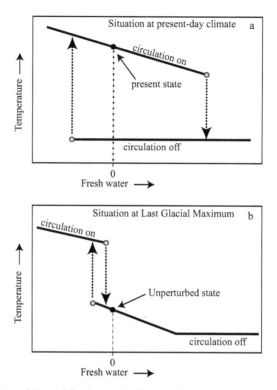

FIGURE 8.10. Stability of the thermo-haline circulation with respect to variation in the freshwater input is thought to have been different in glacial times (bottom panel) as compared to the present-day situation (top panel). (Modified from reference 18.)

large and that the critical threshold is far from the current condition. This fits with the findings that the climate has been remarkably stable over the past 10,000 years compared to the wild excursions found in glacial times. The apparent instability of the glacial climate may in part be due to a different stability picture of the thermo-haline circulation under cold conditions (figure 8.10b). Models suggest that during the last glacial maximum, the normal climate was one in which the circulation was inactive but relatively close to a threshold at which the circulation could be switched on by episodes of reduced freshwater input. This may explain the brief warming episodes known as Dansgaard-Oescher events. Obviously, situations between those two stability pictures may have arisen when the climate slipped from the last ice age into the warm interglacial period in which we live now. It seems plausible that the Younger Dryas happened at a moment when the thermo-haline circulation was switched on but the threshold for a switch-off was still close.

El Niño Southern Oscillation and Other Weather Cycles

The climatic shifts that the thermo-haline circulation can bring about in the Northern Hemisphere happen typically within decades, whereas the perturbed states persist for centuries. Flips between contrasting states happen on faster timescales too. Well known are the irregular oscillations in the Southern Hemisphere between an El Niño and a La Niña phase. This pattern, known as the El Niño Southern Oscillation (ENSO), seems to be related to a delayed feedback mechanism between the Pacific Ocean and the weather.[30] This goes roughly as follows: Warm sea surface temperatures along the eastern equatorial Pacific weaken the westerly winds above the equator. This has two effects. On one hand, it excites a downwelling that deepens the thermocline that separates the warm surface water from the cold deeper water. This dip in the thermocline travels eastward to the South American coast, where it eventually promotes warming of the coastal water—the actual El Niño event. On the other hand, the weakening of the winds initiates an upwelling signal that travels westward, is reflected by the western boundary of the ocean, and on its way back counters the downwelling wave that caused El Niño, terminating the event. Obviously, marine ecosystems are affected by these changes, but the altered patterns of sea surface temperatures also invoke changes in the weather that spread far beyond the Pacific region through so-called *teleconnections*.

Although El Niño events happen roughly every 2 to 5 years, the period between two events is quite irregular, and also the intensity of events varies widely. While some events are so weak that they trigger only moderate weather changes, other so-called *mega El Niños* disrupt the weather pattern tremendously, causing dramatic changes in nature[31] and causing severe problems for societies.[32] The chaotic behavior of the ENSO fluctuations is probably due to stochastic resonance of the

Pacific oscillator with seasonal forcing.[30,33] In fact, this is quite analogous to the irregularity in glacial cycles resulting from stochastic resonance of oscillation in the Earth system with quasi-periodic forcing
from changes in solar irradiation resulting from variations in the
Earth's orbit discussed earlier.

Although the ENSO cycle is perhaps the most obvious and pronounced repetitive pattern in the weather, it is not the only one. Well
known is the North Atlantic Oscillation (NAO), an irregular alternation between contrasting conditions that has large impacts on the
weather and nature in the region. For instance, as it affects the position of dominant pressure systems, it causes a change in dominant
wind direction that affects the patterns of temperature and rain in
Western Europe profoundly. Yet another irregular pendulum in oceanic
and weather conditions that has been discerned is the Pacific Decadal
Oscillation (PDO). The emerging picture is one of an Earth system
that has several intrinsic oscillators that influence each other and are
at the same time affected by the seasonal cycle of solar irradiation, resulting in a chaotic pattern that behaves more or less periodically at
times but that varies considerable in frequency and amplitude.

PROLONGED DROUGHTS AND OTHER RARE EVENTS

While these recurrent chaotic oscillations in the climate system have
attracted much attention, there are also records of incidental prolonged periods of deviating weather patterns reflected in indicators
such as hurricane frequencies, flood regimes, and droughts. A well-
known example is the drought that hit the United States in the 1930s
(figure 8.11). As much of the natural vegetation had been removed for
agriculture, the winds in this exceptionally dry period that lasted almost a decade eroded the dry top layer of the soils, massively accumu-
lating dust in other places. During this dramatic period, known as the
Dust Bowl, over half a million people left the affected areas, and the
already depressed economy suffered deeply. Model simulations suggest that cooler than normal tropical Pacific Ocean surface temperatures combined with warmer tropical Atlantic Ocean temperatures
can create conditions in the atmosphere that invoke such a drought in

FIGURE 8.11. Sand-covered machines at a farm in Dallas, South Dakota, in 1936 illustrate why the persistent drought that hit the United States in the 1930s has become known as the Dust Bowl. (Image from public domain, at *http://en.wikipedia .org/wiki/Image:Dallas_South_Dakota_1936.jpg.*)

the region.[34] Although the Dust Bowl was not characterized by particularly strong ENSO events, it seems not unlikely that extremes in climatic oscillations such as ENSO might swing the climate into such prolonged periods of deviating weather. For example, La Niñas (the opposite phase of El Niño) are marked by cooler than normal tropical Pacific Ocean surface water temperatures, which help creating dry conditions over the Great Plains region of the United States.

More recently, the Sahel region has been affected by a persistent drought that lasted much longer than expected from normal fluctuations in the weather.[35] This drought, which is thought to have led to the death of over a million people, has often been related to the effects of human-induced desertification. However, just as in the Dust Bowl, alteration of the pattern of oceanic temperature distributions appears to have been an important driver.[36] Further back in history, there are indications of multidecadal droughts inducing population dislocations, urban abandonment, and state collapse in the New World as well as the Old World.[32]

In summary, there are numerous indications of episodes of persistent distinct weather patterns that contrast enough with the prevailing conditions to cause serious trouble to societies.

8.4 Synthesis

The emerging picture is one of an Earth system that shows signs of chaotic dynamics and catastrophic transitions on all timescales. The weather was among the first systems that were suggested to be fundamentally chaotic and unpredictable beyond timescales of days.[37] Similarly, chaos rules the dynamics of ENSO fluctuations and other multiyear climatic patterns, while the rhythm and amplitude of glaciations appears to be chaotic on multimillennial scales. Jumping out of the chaos, distinct regime shifts appear on all timescales. Let us walk back in time and summarize a few examples with their suggested drivers:

1. A drought of almost a decade hit the Great Plains, creating the Dust Bowl in the 1930s, invoked by changes in the ocean temperature distribution.
2. About 5,000 years ago, the Sahara desert was suddenly born out of a vegetated region with lakes, a shift invoked by subtle change in insolation resulting from gradual changes in the Earth's orbit (section 1.2).
3. About 13,000 years ago, temperatures in Greenland plummeted more than 10 degrees Celsius, throwing the region back into ice age conditions for centuries until a sharp temperature jump upward ended this Younger Dryas episode. Both changes seemed to result from change in the freshwater flow into the Atlantic Ocean invoking the thermo-haline circulation to shift between alternative stable states.
4. Thirty-four million years ago, the Earth moved from greenhouse to icehouse conditions suddenly and irreversibly. This shift was likely due to reduction in atmospheric carbon levels as CO_2 consumption increased when the Himalayas grew.
5. Fifty-five million years ago, a powerful global warming event turned the Arctic Sea into a 23°C freshwater pool. The likely cause was a runaway process of release of the greenhouse gas

methane from marine sediments as methane hydrates became
unstable when a critical temperature was exceeded.

6. The entire planet switched between a "snowball" state and a very
hot regime several times prior to 600 million years ago. Probably, plate tectonics caused weathering rocks to deplete atmospheric carbon levels far enough to trigger a runaway glaciation,
whereas slow refill of the atmosphere from volcanic sources
eventually triggered the runaway warmings.

Although there is still much uncertainty in such reconstructed scenarios, it seems safe to conclude that most of these apparent regime shifts
appear to be related to gradual changes pushing the system across a
critical threshold. Clearly, studies of past climatic change can hardly
be read without contemplating implications for future developments.
Indeed, many articles in this field end with some speculation in this
direction. Almost invariably, question marks dominate: If gradual
warming triggered methane hydrate instability in the past, can we
expect a similar runaway warming again? If decreasing CO_2 levels
moved the climate from a greenhouse to an icehouse in the past, could
a future increase in CO_2 result in the reverse shift with permanent loss
of ice caps? If warming events in the past were typically sharp compared to cooling, could the current warming speed up too? If freshwater input from melting ice and increased precipitation switched off
the thermo-haline circulation in the past, could it happen again given
the climate change scenarios?

Often, a closer examination suggests that it is unlikely that the
same thing will happen soon again. However, there are no past analogies of the current situation, and we remain quite uncertain so far in
our guesses as to what might happen over the next century. Nevertheless, there are two things on which most of the climate scientists do
agree: (1) We are altering CO_2 dynamics and other aspects of the Earth
system profoundly and increasingly, and (2) the response of the Earth
system to changes is highly nonlinear and numerous thresholds for
small and big changes exist. Global change thus implies an increased
probability of hitting such thresholds sooner or later. While this difficult aspect has barely been addressed in most predictions of climatic
change, it seems obvious that it may be worth focusing on the issue of
abrupt climate change more in the coming decades.

CHAPTER 9

Evolution

9.1 Introduction

Like the evolution of the climate system, the evolution of life on our planet has been a bumpy process. In fact, the two are intimately linked. For instance, the evolution of photosynthesis may eventually have led to a "snowball Earth" glaciation, terminating nearly all life on the Earth. Subsequently, the turbulent end of this episode, resulting in 50°C oceans for some time, marked the beginning of the Ediacaran and Cambrian explosions of new life-forms. Certainly, one of the most intriguing themes in evolution theory is the idea of *punctuated equilibrium*.[1] It suggests that evolution is characterized by long periods of little change (stasis) punctuated by bursts of intense evolution and speciation. Most of the argument has been based on studies of the always imperfect fossil record. However, analysis of gene-sequences lends independent support to the idea that a substantial part of evolutionary change is the result of brief episodes of rapid evolution.[2] What might cause such a pattern?

On a long timescale, life on the Earth has been repeatedly shaken up dramatically by the occurrence of mass extinction events. Arbitrarily, five mass extinctions are distinguished, and various smaller ones exist (figure 9.1).[3] After each mass extinction, evolution compensates for the loss of species, to reach similar or higher levels of diversity in about 10 million years.[4] Often, massive extinction episodes allow

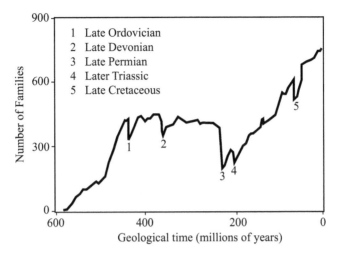

FIGURE 9.1. Diversity through geological time of marine animals. The "big five" mass extinctions are visible as abrupt drops in the diversity curve. (From reference 3.)

groups that existed before to rise to dominance and differentiate.[5] Although a lot of the turnover of species has happened more gradually,[6] the overall picture certainly remains a jerky one, with distinct episodes of intense extinction and speciation (figure 9.1). One could argue that in general, punctuated dynamics in evolution may have been caused at least in part by external perturbations. Certainly, episodes of rapid climate change and incidental disasters such as meteor impacts and volcanic eruptions appear to have had large effects.

On the other hand, an intriguing question is whether some of the marked species turnover events could represent self-propagating dynamics that merely need some trigger to start rolling. Some theoretical work suggests that such dynamics could plausibly occur. One suggestion is that evolution might be characterized by self-organized criticality (section 4.1). This implies that the evolutionary process by itself would every once in a while produce a critical unstable state at which small or large avalanches of change occur.[7] As explained earlier, the evolutionary record suggests that the size distribution of major extinction events follows a power law that is consistent with what would be expected from self-organized criticality but is not a proof that such a process has been at work. Another possible view is that gradual environmental change tips the balance when a critical point is passed, and life on the Earth reorganizes into another stable end-point community.[8] (See the theory on such alternative stable states in section 4.2.) Also, on a more specific level, it has been argued that positive feedbacks may have been responsible for runaway dynamics in many of the most important and spectacular processes in evolution, including the evolution of sex and genetic systems, mating systems, life histories, and complex cooperation in insects and humans.[9]

Before talking more about different theories of such changes, I invite you to take a look at some of the most profound events of evolutionary change. Subsequently, we will ponder which of the generic mechanisms of critical transition (if any) might be involved in causing the apparent jerky nature of evolution. We start at the Cambrian explosion, at which the body plans of most, if not all, modern life-forms came into existence, and proceed from deep time to more recent periods, when an apparent global warming event triggered the evolution of primates and other modern mammals. Again, a cautionary note is in order when it comes to the scientific certainty in this field. Just like the study of ancient climate change, the reconstruction of evolutionary history reads like a detective tale. We have to rely on scarce clues, most events are unique rather than repeated, and experimental approaches on meaningful scales are out of the question. Even if we get close to determining the dynamics of evolution, unraveling the mechanisms involved is a daunting task.

9.2 Early Animal Evolution and the Cambrian Explosion

Looking at the big picture of the history of life, one of the most puzzling things to me is that for most of the nearly 4 billion years that life has existed on the Earth, evolution produced little beyond bacteria and algae. This long period came to an abrupt end about 600 million years ago, with what is perhaps the most stunning period of innovation in the history of evolution: a sudden radiation of life-forms known as the Cambrian explosion. Scientists have long been puzzled by its abruptness and the apparent lack of obvious predecessors to the Cambrian fauna. The change is spectacularly visible in several rocks around the world that represent a time-series of sediments originating around this period. Starting at the bottom, the older layers show little remains of life at all, until all of a sudden layers full of distinct and varied fossils appear. This phenomenon has been known for more than a century, and its abrupt character bothered Darwin as one of the principal objections that could be raised against his theory of gradual evolution by natural selection.

A closer look reveals that there have really been two early "explosions" of life-forms: first the so-called Ediacaran, and "only" about 20 million years later, the famous Cambrian. The dynamics and possible causes of those early evolutionary revolutions are much more difficult to reconstruct than the relatively recent events such as the rise and fall of the dinosaurs. However, an insightful overview of what is known and how it might be interpreted has been sketched by Andrew Knoll and Sean Carroll. I follow their reasoning, and references can be found in their review.[10]

The first breakaway from the billions of years of largely microbial life was a rapid rise and fall of mysterious creatures known as the Ediacaran fauna. These include simple blobs, but also a range of forms built of repeated tubelike units, worms, and other soft-bodied forms. In addition, there are numerous fossils of traces, probably made by wormlike animals slithering over mud. Much remains unclear about the precise nature of this first burst of elaborate life-forms. However, most of the Ediacaran fauna apparently did not come to stay. A few of the forms seem to have persisted into the next period known as the Cambrian, but the majority disappeared suddenly without leaving obvious descendants. The collapse of the Ediacaran fauna is followed by the spectacular radiation of novel life-forms (figure 9.2) known as the Cambrian explosion. All of the main body plans that we know now evolved in as little as about 10 million years. It has long been thought that this apparent explosion of diversity might be an artifact. For instance, it could be that earlier rocks were not as good for preserving fossils. However, very well preserved fossils do exist from earlier periods,

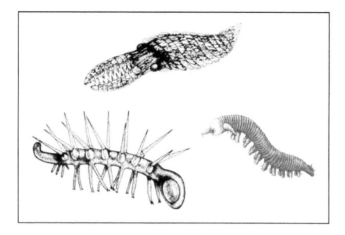

FIGURE 9.2. Some of the multitude of bizarre new life-forms that occurred suddenly in the Cambrian period. (Drawings by Mary Parrish, courtesy of the Smithsonian Institution.)

and it is now generally accepted that the Cambrian explosion was real. The big questions then are why did so many life-forms evolve so rapidly, and why at that particular moment?

One idea is that a rather elaborate genetic toolkit is needed to produce the bilaterian body plans that set much of the Cambrian fauna apart from the largely radially symmetric or threefold symmetric earlier animals. Arthropods are among the most abundant complex Cambrian fossils. Their modern descendants include animals such as crabs, shrimps, myriapods, and insects. Segment and appendix differentiation in these groups is regulated by a particular set of genes known as *Hox* genes. It has been suggested that such complex sets of genes needed to evolve in order to allow the differentiation. However, it appears that essentially the same set of *Hox* genes is used across all animals of the modern arthropod (as well as the onychophoran) clade. This tells us that the genes must have been there before those groups separated. Apparently, the Cambrian and modern diversity evolved around an ancient set of *Hox* genes that was already in place before the Cambrian explosion.

So if development of the genetic toolkit by itself was not the trigger of the Cambrian explosion, what else can explain it? Some environmental factor perhaps? Indeed, more may have been involved in getting the chain reactions of the explosive Ediacaran and Cambrian evolutions going. As mentioned earlier, one or more snowball Earth glaciations probably froze the Earth, while extreme warming between them heated the oceans to temperatures of about 50°C (section 8.1). The last of those major Proterozoic ice ages just predates the Ediacaran radiation. Models and phytoplankton extinctions suggest that the snowball Earth climatic dynamics events must have had a severe impact on the biota and probably put the brakes on early animal evolution before this period of turmoil was over. Or perhaps, this was a shake-up allowing the subsequent Ediacaran burst of evolution?

So what about the end of the Ediacaran life-forms and the start of the Cambrian explosion then? Again, there are indications that a massive environmental perturbation happened at this point. Around the world, sediments reveal a strong but short-lived negative excursion in the carbon-isotopic composition of surface seawater just at the breakpoint between Ediacaran fossil assemblages and those that document the beginning of the Cambrian explosion. The causes of this pattern remain uncertain, but the only comparable events in more recent Earth history coincide with mass extinctions such as the end-Permian crisis, discussed in the next section, when about 95% of the species disappeared. Thus, although reconstructions remain problematic, it may well be that the Cambrian explosion could happen only after an environmental shock "cleared out space" by eliminating much of the Ediacaran fauna.

Once the Cambrian diversification started, interactions among organisms must have further catalyzed diversification. For instance, predation may have promoted the evolution of skeletonized animals. Also, algae that had been around for a long time started diversifying spectacularly during the Cambrian explosion. This suggests that the Cambrian explosion was an ecosystem-wide phenomenon. Phrased in another way, selective forces were no longer predominantly abiotic. Instead, biological components of the environment became principal features shaping the fitness landscapes, as they are now. The heart of the great radiation of forms may thus have been a positive feedback in which diversity of organisms promoted diversity of niches, which further promoted diversification of life-forms.

9.3 The End-Permian Extinction

Life did not proceed smoothly after the Cambrian explosion. Three very large and several smaller extinction events hit the biota in the following hundreds of millions of years. Most of them pale in comparision to an event 251 million years ago, at which an estimated 95% of all species on the Earth became extinct. Although this mass extinction was of a stunning magnitude, it has been difficult to reconstruct what happened until recently, when several pieces of the puzzle started fitting together. The scenario I describe here is based on a review by Michael Benton, where further references can be found.[11]

The few continuous fossil-bearing rock sections that document the period around this mass extinction show that the latest Permian seas were full of life. The sediments have burrows made by a range of benthic animals living, feeding, and moving through the sediment. The communities were apparently diverse and complex. In contrast, sediments deposited immediately after the extinction event largely lack burrows, and those that do occur are very small. Fossils of marine benthic invertebrates are extremely rare. These observations, together with geochemical evidence, suggest that the dramatic change in marine life (figure 9.3) may have been caused by widespread anoxia. Terrestrial life too was very diverse before the disaster. Amphibian and reptile faunas had reached high levels of complexity, and numerous groups of plants provided a diversity of habitats. Massive species losses on land happened at the same time that the marine changes occurred. Few species appear to have survived, and it seems that soils were washed off the land completely.

The question of what caused this apparent disaster is not so easily answered. Mass volcanism seems to have played a major role. The chemistry of deposits from this period also gives some interesting clues about the nature of the environmental changes. Just below the Permo-Triassic boundary, there is a sharp shift in oxygen isotope values indicating a large global temperature rise. Such global warming can reduce ocean circulation and limit the amount of dissolved oxygen, explaining a widespread anoxia at the marine sediment surface. What may have caused such warming? Volcanic eruptions in what is now Siberia were massive, covering a surface as large as the current

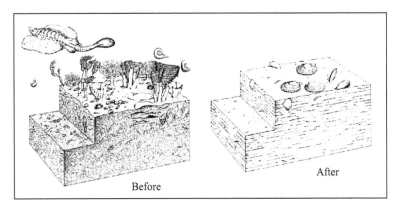

After

Before

FIGURE 9.3. An impression of marine life before and after the mass extinction event at the end of the Permian period. (From reference 11.)

European Union with lava. However, by themselves, they may not have been enough to explain the climatic change. A clue in this sense is that studies of carbon isotopes from that period suggest a strong increase in the light carbon isotope. This would fit with methane released massively from buried gas hydrates. As explained later, the same mechanism most likely caused the more recent thermal maximum at which evolution of primates, deer, and horses started. The idea in the case of the end-Permian warming is that initial global warming, caused by the Siberian eruptions, melted frozen gas hydrate bodies. This input of methane into the atmosphere caused more warming, melting further gas hydrate reservoirs, and leading to a runaway warming process (see also section 8.1). Spiraling out of control, this global warming may have been part of the explanation of why for the second time after the snowball Earth episode, life came close to complete annihilation 251 million years ago. Only an estimated 5% of the species survived as the basis to rebuild life. It took about 100 million years for global biodiversity to return to preextinction levels (although it appears that recovery of ecosystem structure was about ten times faster than that). On land, for millions of years, virtually the only four-legged animal surviving was the plant-eating Lystrosaurus, subsisting on the few surviving herbaceous plants. Forest communities were absent for many millions of years. Although the precise mechanisms are still not well resolved,[12] all this suggests that life was very tough after the apocalyptic events. Rather than a subtle trigger for innovation, this was a setback that took a long time to overcome.

9.4 The Angiosperm Radiation

Entering more recent geological times, one of the greatest radiations of terrestrial species is the rise of flowering plants (*angiosperms*) in the mid-Cretaceous period. The major groups originated relatively early, starting about 130 million years ago. Then, after some time delay, a sharp rise to ecological dominance occurred 100 million to 70 million years ago. It is another swift change in evolution (together with the Cambrian explosion) that puzzled Darwin. Interestingly, unlike in the other examples of evolutionary transitions I have presented, there is no indication for a major extinction event or environmental disturbance in this period. Attempts to explain the sudden massive angiosperm diversification have stressed the importance of key innovations such as pollination by animals, seed dispersal by animals, and life history flexibility. However, there does not seem to be a simple clue to the radiation in terms of biological features that were the key to sudden success.[13]

Another line of work has tried to look into the ecology of early angiosperms. This produced a range of hypotheses, including the following characterizations:

- Weedy shrubs that lived in open, disturbed habitats of semiarid tropical to subtropical regions[14]
- Plants of disturbed streamside habitats in mesic environments[15]
- Fast-growing, semiherbaceous plants of sunny, unstable streamsides that tolerated disturbance and had high leaf photosynthetic capacity and short generation times[16]
- Woody plants that grew in dimly lit, disturbed forest understory habitats and/or shady streamside settings[17]

Note that the only overlapping aspect of these descriptions is that early angiosperms were characteristic of disturbed sites. This suggests that they had difficulties competing with earlier vascular plants such as ferns and the abundant early seed-bearing groups (related to, for example, conifers) collectively known as gymnosperms. How then did they at some point take off to escape from this subordinate position and become dominant around the globe?

The Dutch vegetation ecologist Frank Berendse suggested a fascinating explanation to me. His idea is that angiosperms need higher nutrient levels than the gymnosperms that dominated before, whereas at the same time, angiosperms promote soil nutrient levels by producing leaf litter that is more readily decomposed. This implies a positive feedback that might produce a runaway process once angiosperms reach certain abundance. The observation that early angiosperms occurred largely on disturbed sites would be well in line with this suggestion, as those sites might be richer in nutrients than the places dominated by gymnosperms with their poor leaf litter.

This hypothesis fits especially well with modern observations on critical transitions from gymnosperms to angiosperms such as described for raised bogs (section 11.4). We could imagine that the same competition mechanism might apply as suggested for maintenance of dominance by peat mosses (which are also gymnosperms) in bogs. Suppose that like modern gymnosperms, the ancient gymnosperms could thrive at low nitrogen concentrations but were also keeping nitrogen contents in the soil low, because of the kind of leaf litter they produced. This would then lead to alternative stable states in the competition, where angiosperms occurring at a critical nitrogen concentration of the soil would be exceeded (figure 9.4). This might thus explain why the first angiosperms could enter only at disturbed sites. However, once angiosperms are able to take over, they promote higher nitrogen levels in the soil, thereby securing their own competitive dominance. From their strongholds, seed dispersion might then have helped to gradually take over adjacent habitats previously dominated by gymnosperms. Spatial models of systems of competition with alternative states show that such domino effects can indeed be expected.[18] One can thus imagine a gymnosperm-dominated world in which angiosperms have been suppressed for a long time. However, once unleashed, they would become dominant in a runaway process.

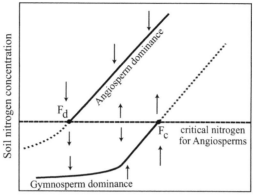

FIGURE 9.4. Graphical model of competition between early angiosperms and gymnosperms.

9.5 From Dinosaurs to Mammals

Perhaps the best-known example of a dramatic shift in nature is the extinction of the dinosaurs at the end of the Cretaceous period (65 million years ago). A browse through any popular encyclopedia for dinophiles is enough to see that the dominance and diversity of this animal group has been truly impressive. It is difficult to determine how many dinosaurs may have actually existed. However, it has been estimated that we now know about less than one third of the dinosaur species, and excluding the avian species, about 1,850 species may have existed.[19] This diversity makes the mass extinction particularly puzzling. How can a species turnover be so radical as to eliminate virtually the entire group?

It is widely accepted that a huge asteroid impact must have been an important trigger of the extinction. However, discussion remains on questions such as whether volcanic erruption may have played a major role too.[20] Also, there is debate as to whether the extinction was abrupt[21] or gradual and explained by cooling atmospheres and the spread of temperate floras favoring mammals over dinosaurs.[22] Certainly, it was a period of environmental turmoil, and the dinosaurs were not the only group that was affected. There was a major turnover of fauna at this event. The change was so large that it is considered to separate two distinct periods in the fossil record: the Cretaceous and the Paleocene. The big picture suggests that at the Cretaceous–Tertiary (or K/T) extinction event, about 50% of all species became extinct. The K/T extinction was rather uneven. Some groups of organisms became extinct, some suffered heavy losses, and some appear to have fared relatively well.

From a human viewpoint, it is significant that this extinction event seems to have facilitated the start of the radiation of mammals. Although mammals coexisted for a long time with the dinosaurs, they remained relatively unimportant until the K/T event, at which point, they started spreading to great diversity and abundance. Thus, just as in the case of angiosperms and the Cambrian explosion, it seems that the innovation was there millions of years before the true radiation.

9.6 Global Warming and the Birth of Primates, Deer, and Horses

Most modern orders of mammals appeared about 10 million years after the end of the Cretaceous period at the so-called Paleocene–Eocene boundary about 55 to 55.5 million years ago. One of the surprising results of high-resolution studies of mammals from this transition is that the modern orders of deer, horses, and primates actually appeared together suddenly, without any known precursors at this time. This is particularly surprising because they are not closely related. Another surprising result is that most of the earliest species were significantly smaller than their immediate descendants, while at the same time, some other groups living at this time appear to be dwarfed. In a nice overview, Philip Gingerich[23] shows that this may be related to a catastrophic warming episode that occurred when the methane buffers in the ocean floor became unstable, much like the warming that may have killed so many species at the end of the Permian. As mentioned earlier, this is one of the examples of runaway processes that appear to have caused critical transitions in climate history.

Although the impact of this warming was small compared to what happened at the end of the Permian period, it is not surprising that a

global greenhouse event of this magnitude had evolutionary consequences. Indeed, the synchronous appearance of deer, horses, and primates did not stand alone in this sense.[23] There were also extinctions.
For instance, about half of benthic foraminifera in the oceans became
extinct, probably because of the reduction in dissolved oxygen resulting from the higher water temperatures and reduced circulation. On
land, some animals that were characteristic of Paleocene faunas, such
as the crocodile-like reptile Champsosaurus and a primate-like mammal Plesiadapis, became extinct. In those more recent evolutionary
events, we are able to reconstruct the changes more precisely than we
can for events such as the Cambrian explosion. One of the aspects that
emerged in the analyses of this episode is that the climatic change as
well as the apparent evolutionary responses happened in only thousands of years. This contrasts strongly to the million year timescales
we usually associate with evolution.

9.7 In Search of the Big Picture

Clearly evolutionary history is a dazzling field, full of rich and idio-
syncratic details. Also, the major transitions I have highlighted differ
in essential aspects. Nonetheless, there seem to be some general
patterns:

- *Explosive radiation of new groups of species is often triggered by an
  environmental perturbation.* Events such as the Cambrian explosion, the radiation of mammals, and the synchronous appearance
  of primates, deer, and horses appear to have been triggered by environmental disturbance. Although the angiosperm radiation does
  not seem to be initiated by a major disturbance, angiosperms appeared first on locally disturbed sites.
- *Novel features tend to be there for a considerable time before explosive radiation takes place.* For instance, mammals coexisted with
  dinosaurs for a long time before they took over, the main groups
  of angiosperms evolved well before the great radiation, and *Hox*
  genes allowing body plan differentiation probably existed well
  before the Cambrian explosion.

- *The maximum speed of evolution seems very fast compared to the average rate of change.* This is apparent from the warming-induced evolutionary dynamics of modern mammals as well as many studies of present-day evolution. Although earlier evolutionary changes may have been equally fast, only high-resolution analysis limited to more recent events may be able to reveal the true maximum speed.

Taken together, this suggests that developed communities tended to suppress major radiations of new life-forms. It appears as if the Ediacaran fauna suppressed Cambrian fauna, gymnosperms suppressed angiosperms, dinosaurs suppressed mammals, and early mammals suppressed the modern groups of primates, deer, and horses. Although major radiations may have been suppressed, evidence suggests that innovation happened in the "shade" of the dominant communities. So called *stem groups* seem to have developed typically long before the explosive radiation of *crown groups* (figure 9.5), and the novel life-forms may often have existed for millions of years playing only a minor ecological role.

It thus seems as if the ecological niches tended to be well occupied by the existing community. Not only may the species have evolved to fill the niches almost hermetically, the niches themselves were also shaped by the existing biota. Surely, since the evolution of complex animal life-forms after the Proterozoic, biotic components such as predators, parasites, symbionts, prey, and competitors have been major determinants of ecological niches. Would it have been impossible to take over dominance in some of those well-tuned and coevolved systems without the help of some major disturbance to their structure? And how then did novelty form, preparing the system for a takeover? Obviously, ecosystems are never "hermetically" closed. There are always "leftovers" and disturbances that clear out spots where competition and predation may be temporarily less severe. Perhaps this provided enough space for innovation without triggering the major critical transition.

This may seem like a rather loose reasoning. However, mathematical models provide some evidence that this kind of critical transition between complex alternative stable communities may indeed be

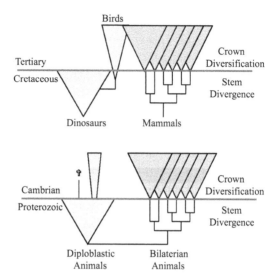

FIGURE 9.5. Both the radiation of mammals and the Cambrian explosion of bilaterian animals probably happened after environmental perturbation of the previous faunas that then became almost completely replaced. (From reference 10.)

reasonably expected. As discussed earlier (section 4.2), models of a multitude of hypothetical species tend to settle to one of several possible stable endpoint communities.[24] Such endpoint communities cannot be invaded by any of the species from a large pool of other hypothetical species, unless something happens to the environment. There are two kinds of environmental change that may cause a shift.[8] First, a major disturbance, reducing population densities of many of the species, can tip the system into an alternative basin of attraction, where it evolves toward a new stable community. Second, slow environmental change may reduce the ecological resilience of the dominant community. This reduces the size of the basin of attraction around the existing community, eventually allowing even a small disturbance to trigger a shift to an alternative basin of attraction. In fact, if resilience of the existing community has become small enough, a critical transition in the community may even be triggered by inconspicuous vagaries of the environment (figure 9.6).

The phenomenon of alternative endpoint communities has been demonstrated experimentally on small scales.[25] In a wider sense, evi-

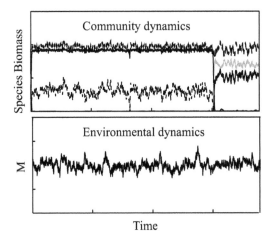

FIGURE 9.6. Simulation illustrating how business-as-usual fluctuations may trigger a shift to an alternative community of hypothetical interacting species in a situation where the resilience of the community has become low. (Modified from reference 8.)

dence for loss of resilience and shifts between alternative stable states has been accumulated for various contemporary ecosystems. Scales vary from the size of patches of kelp forest,[26] lakes,[27] and reefs[28] to critical transitions on a regional scale such as the shift from a vegetated to a desert state in the Sahara about 5000 years ago.[29] One view of the major evolutionary transitions may thus be that they represent similar reorganizations, only on a massive scale.

Clearly, one cannot simply jump from ecological timescales to evolutionary ones, where species can to a certain extent evolve in response to environmental change. So far, the only theoretical work suggesting that evolution may lead to discontinuous jumps are the self-organized criticality models.[7] These produce patterns much like the evolutionary record.[30] However, the model is a gross caricature of true evolution. Also, the fact that it produces patterns as observed in nature (a power law of extinction avalanche sizes) is no proof that it does so for the right reasons. For instance, the same patterns may arise from a model without coevolutionary processes in which environmental perturbations are the main drivers of extinction.[31] In fact, the evidence for environmental drivers of major change in evolution is so pervasive that it would be stubborn to maintain that purely intrinsically generated instability may explain the jerky nature of evolutionary change.

The more interesting question may be whether instead of being purely intrinsically generated, or purely externally driven, major evolutionary changes may have resulted from runaway processes driving species assemblages into an alternative basin of attraction upon perturbation. Obviously, the case for such an explanation is not built as easily as for small contemporary systems. However, some of the possible indications that a shift to an alternative basin of attraction is involved (see chapter 14) may be checked even against the scarce data we have. Let us first summarize the alternative explanations for events of major evolutionary turnover that we should consider. Roughly, four explanations can be distinguished, as follows:

- A key innovation allows a new group to be come abundant and outcompete the previously dominant species.
- Evolution drives the community to a self-organized critical state, causing an avalanche of extinctions.
- A major disturbance eradicates the previously dominant group, allowing a new group to take over.
- A stepwise persistent change in environmental conditions leads to a radical change in species.

The first two explanations assume intrinsic drivers of change (evolution itself); the last two assume external causes. Pondering the evidence we have, the first explanation seems inadequate. It may well drive much of the smoother evolutionary change.[6] However, in most of the highlighted cases, the innovative features were present in stem groups, long before the explosive takeover. The second explanation might well have an element of truth, but as argued, there is little evidence, and it certainly downplays the role of pervasive perturbations in the fossil record. The third explanation seems attractive but is clearly overly simplistic too. Rather than specifically killing all organisms of a certain group, most disturbances seem to have wreaked havoc on life in a messier way. Most groups often suffered, and the outcome of the struggle for life afterward seems to have been fairly unpredictable *a priori* in most cases.[6] The fourth explanation may play a role in early animal evolution when oxygen availability suddenly allowed large body sizes. However, in most cases of major evolutionary turnover, we have no evidence that there was a marked contrast between environmental conditions before and after that explains the difference between the new and the old communities.

From a dynamical systems theory point of view, a logical alternative explanation for some of the major evolutionary transitions is this:

- Environmental and/or evolutionary change has reduced resilience of the dominating community, allowing a disturbance to trigger a runaway shift to an alternative stable community.

This could explain why sharp evolutionary transitions are often related to large perturbations but can also happen without obvious massive disturbance. The latter could happen if evolutionary change or environmental change has reduced resilience of the current state to nil (figure 9.7). The big underlying idea is that the composition of the Earth's biota has alternative basins of attraction, just like the stable endpoint communities demonstrated experimentally and in models of simpler communities on smaller timescales. Although this may seem an attractive alternative to the first four explanations of evolutionary shifts, "proving" such a hypothesis is obviously not so easy.

In ecology, mechanistic models play an important role in making the case for alternative stable states. For the angiosperm radiation, a feedback on nutrient availability might explain alternative basins of attraction. However, mechanisms that generate alternative stable states in complex communities can be difficult to unravel even in models,[8] and there is little hope that we would do better in unraveling such mechanisms in most past communities.

Empirical evidence for alternative basins of attraction (chapter 14) includes the observation that the system may not return to the same state upon perturbation (as observed in the fossil record). A more subtle clue may be that in the course of the transition, there may be a phase of acceleration as the system is pushed away from the unstable watershed between the alternative basins of attraction. Of course, such subtleties may be less easily identified upon a massive perturbation than

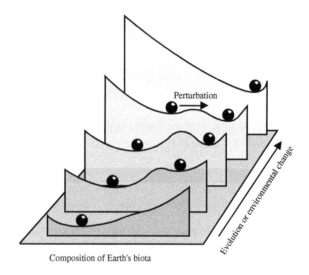

FIGURE 9.7. A view of evolutionary transitions as stability shifts, triggered by environmental perturbations. Slow evolutionary or environmental change may eventually destabilize the current state, allowing increasingly small perturbations to trigger a transition.

if the system is just tipping over a critical point. If a new group of species replaces an old one, there should be such an accelerating episode decline of the old species, as well as the rise of the new life-forms. It would be interesting to search the fossil record for this and other clues that part of the drastic reorganizations may be interpreted as critical transitions. In fact, in the face of the estimated impact of global change on species extinction rates and environmental conditions, this question may be of more than academic interest.

9.8 Synthesis

Putting all lines of evidence together, we are quite sure that periods of relatively little change have been interrupted by relatively brief episodes of spectacular evolutionary development. Unfortunately, it is very difficult to unravel the mechanisms behind this pattern. Certainly, major disturbances such as periods of intense volcanism and meteor impacts have had their impact. However, as I have argued, it is not unlikely that some of the major evolutionary turnovers may have been critical transitions in the sense that they were characterized by self-propagating runaway change once a critical threshold was passed. It will be a challenge to tease out the hallmarks of such change from the evolutionary records.

CHAPTER 10
Oceans

As you saw in the chapter on climate (chapter 8), the oceans are implicated in several examples of profound and abrupt climatic change. However, looking in more detail, ocean ecosystems are subject to occasional sharp regime shifts too. The scale of those changes varies from individual reefs and estuaries to synchronized regime shifts that affect an entire ocean basin. While lakes are close to ideal objects for ecosystem studies, unraveling the functioning of ocean ecosystems is a formidable challenge. Even just determining what kinds of fish and planktonic organisms occur in a region and what their population densities are appears almost impossible, and far more challenging than that is determining what drives apparent trends and shifts in these ecosystems. In lakes, we can experimentally eliminate fish or add nutrients and look at the effects on a whole-ecosystem scale. Such experiments are impossible on an ocean-basin scale. Certainly, overfishing, coastal eutrophication, and climatic swings have large impacts and may be interpreted as kinds of experiments, but doubts about the causal relationships remain necessarily more pronounced than in most experimental lake studies. As a result, compared to lake studies, science for the ocean has been focused more on the challenge of detecting interesting patterns such as regime shifts and has advanced less in explaining those patterns. Somewhat easier to study than the open ocean are coastal systems such as kelp forests, coral reefs, and estuaries. These are easier to sample and even allow enclosure experiments to help unravel some mechanisms. Nonetheless, compared to small lakes that can be manipulated entirely, it remains more difficult to understand the forces that drive the dynamics of these open marine systems. What follows is an overview of sudden shifts that have been studied in marine ecosystems, together with the state-of-the art speculations about the mechanisms that may be involved.

10.1 Open Ocean Regime Shifts

While studying patterns in the data on fish catches, zooplankton abundance, shellfish condition, water color, temperature, and numerous other aspects, scientists started realizing in the early 1990s that many of those variables occasionally change sharply and in synchrony over large regions. They coined the term *regime shifts* for such revolutions.[1,2] Because of the mere challenge of showing patterns in the tremendous amount of noisy and spatially spread data, it took some time to convince the oceanographic community that this was something significantly different from the usual blur of permanent change.[3] However, the existence of marked regime shifts is now well accepted, and the topic has quickly become one of the more intensely studied themes in marine science.[4]

Shifts in the North Pacific

Perhaps the most important impulse for the interest in marine regime shifts came with the discovery of an apparently massive shift in the ecosystem of the North Pacific around 1976–1977.[2,5] Analysis of numerous biotic and abiotic time-series indicated that many variables shifted in conspicuous synchrony at this moment and again (to a different state) around 1988–1989. One way to illustrate this is to normalize all the time-series (to let them run between −1 and 1), subsequently reverse the sign of part of them to make them all flip in the same direction at the moment of the supposed regime shift, and then plot the average value of the resulting transformed series over time (figure 10.1). To check whether the year of the regime shift is really exceptional, the procedure is repeated with the assumed moment of a regime shift in each of the other years for comparison. Later work has

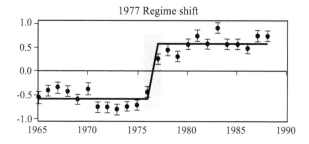

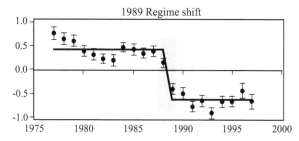

Figure 10.1. Distinct regime shifts occurred in the Northern Pacific Ocean ecosystem around 1976–1977 and 1988–1989. The compound indices of ecosystem state are obtained by averaging 31 climatic and 69 biological normalized time-series. (Modified from reference 2.)

shown that this procedure might give false positives in strongly auto-correlated time-series such as these marine ones. However, using a range of other approaches, the conclusion that the shifts do indeed represent significant discontinuities in the dynamics of the North Pacific ecosystem persists.[3]

Such massive shifts may have implications for regulatory processes such as oceanic CO_2 uptake but may also be directly felt by fisheries. Reproduction and growth of some species may dwindle at a regime shift, while other species fare well. Clearly, fisheries regulations should ideally respond swiftly to such change, for instance, by closing fisheries on some species and relaxing quotas for other species. However, it remains challenging to detect a regime shift in an early phase, and we are far from able to predict in advance when an oceanic regime shift will occur.

The first step beyond the mere problem of detection is of course to find out what drives oceanic regime shifts. Most researchers agree that changes in the oceanic circulation and related weather patterns are the most likely triggers. Indeed, some marked changes in physical patterns occurred at the time of the Pacific regime shifts. Since we also know that circulation patterns affect distributions of temperature and nutrient levels, it is obvious that such changes may be of great importance to numerous biological processes.

So are regime shifts basically physical phenomena, reflected in the biology? Indeed, the dominant view of ocean ecosystem regulation has long been that they are driven overwhelmingly by physical dynamics. However, there are some indications that such a view is too simple. One indication that biology does more that simply track physical dynamics is the fact that shifts from one persistent regime to another are clearly seen in biological variables but not in the physical indicators.[2] This suggests that biotic feedbacks could be stabilizing the community in a certain state, even if shifts to a different state are triggered by physical events. Such a view of an ecosystem with alternative attractors caused by ecological feedbacks is also consistent with the outcome that abiotic conditions appear to overlap between the two different regimes in the North Pacific, whereas the biotic community stayed distinctly altered after the regime shift (figure 10.2).

Other evidence for the idea that the biological communities in the Pacific are not simply tracking environmental fluctuation comes from sophisticated time-series analyses.[6] These reveal a marked distinction between the pattern of fluctuations in physical variables and the patterns observed in biological variables. Physical variables such sea surface temperature appear to fluctuate basically at random. Of course, the series are autocorrelated (today looks like yesterday) as changes in temperature and currents are slow. Such slow fluctuations are best described as "red noise." In this sense, the underwater climate in the ocean differs from fluctuations in conditions in terrestrial ecosystems where vagaries of the weather are more rapid (noise is closer to "white"), and populations should have more difficulties in following such rapid changes, especially if their intrinsic growth rates are low.[7] However, despite the fact that marine organisms might follow fluctuations in temperature and other factors more closely than terrestrial organisms, the same techniques of time-series analysis suggest that the biological communities in the Pacific do something markedly different than just tracking the random environmental fluctuations.[6] In

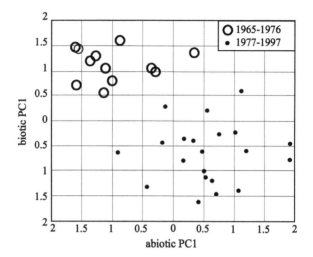

FIGURE 10.2. A plot of the multivariate indices for the overall state of the biotic versus the abiotic system in the Northern Pacific time-series shows that the regimes before and after the 1977 shift differ quite consistently for organisms, whereas they overlap with respect to the physical–chemical conditions. (Modified from reference 3.)

contrast to the dynamics of the physical indicators, the biological time-series appear to have *low-dimensional nonlinear signatures*—that is, they seem to be shaped by feedbacks that lead to nonlinear amplification of the stochastic physical forcing around certain thresholds.

REGIME SHIFTS IN THE ATLANTIC

Although the regime shifts in the North Pacific have attracted much attention, they are certainly not the first observations of marked shifts in marine systems. A famous example of surprisingly sharp changes is the so-called *Russell cycle* (figure 10.3). Rather than a true ongoing cycle, this is a shift in the biological community of the English Channel near Plymouth that occurred in the early 1930s and reverted in the 1960s.[8] As in other regime shifts, several changes occurred simultaneously. Overall catches of young fish dropped, and the herring fishery, which had been a regular winter feature in the region, declined sharply. At the same time, there was a marked decline in the quantity of

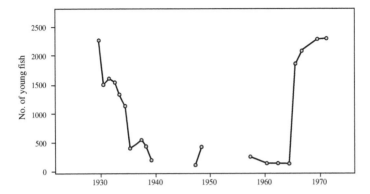

FIGURE 10.3. Changes in the number of young fish caught in standard sampling hauls mark one aspect of the many synchronous changes in the community of the English Channel in the 1930s and 1960s known as the Russell cycle. (Redrawn after Russell et al.; see reference 8.)

zooplankton, and pilchard became more abundant. In the 1960s, a reverse shift appears to have occurred, accompanied by the start of a remarkably productive period for cod and other gadoids. Russell and his coauthors noted that their research area near Plymouth (their research station) is situated on a well-marked boundary between a boreal community and a warm water community and hypothesized that the shifts they described were caused by changes in temperature and circulation patterns that made this boundary move up and down.

About twenty years later, another large-scale shift was noted in the North Sea ecosystem. In the mid-1980s, characteristics of communities of phytoplankton, zooplankton, and fish in the North Sea all appear to have changed in synchrony.[9] A good starting point to examine change in this region is to look at the data collected by the so-called *continuous plankton recorder* (CPR). This is a device that has been towed behind volunteer merchant ships on a monthly basis since 1946. It provides a simple but impressively long and quite consistent series of samples from the upper layer of the water column. The basic indicator used to characterize phytoplankton is the color observed in a standardized way on the piece of cloth that serves as a filter in the device. In the mid-1980s, a marked change in the phytoplankton community occurred (figure 10.4). Overall, phytoplankton biomass

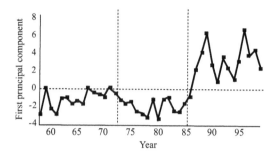

FIGURE 10.4. A shift in the phytoplankton community in the North Sea in the mid 1980s reflected by change in the first principal component of the seasonal pattern of color recorded in samples taken with a continuous plankton recorder. (From reference 9.)

appears to have increased considerably, leaving the samples more intensely colored than in the decades before the shift.

Around the same time, a similarly dramatic shift occurred in the zooplankton community (figures 10.5 and 10.6). Much attention has been focused on a group known as calanoid copepods that can be very abundant and serve as important food for many juvenile fish. The North Sea copepods shifted from a community dominated by large-bodied species of colder waters (for example, *Calanus finmarchicus*) to a community with smaller species (for example, *Calanus helgolandicus*) characteristic of warmer waters. While *finmarchicus* used to occur in spring, *helgolandicus* peaks only in late summer.

Last but not least, the 1980s North Sea regime shift seems to be associated with a drop in the recruitment success of cod and gadoid species (figure 10.6). Mechanisms are difficult to infer from the timeseries. However, it has been suggested that the change in the zooplankton community may have contributed to this change. This is because large calanoid copepods are an important food source for juvenile cod in spring. The overall increase in temperature around the same year might have aggravated the situation for cod, as it implies higher metabolic costs in this critical first stage of life.[9]

The question remains whether the North Sea regime shift may be seen as a shift to an alternative basin of attraction. If one looks at the problem on a larger geographical scale, it appears that the distribution of communities in the entire North Atlantic has changed profoundly

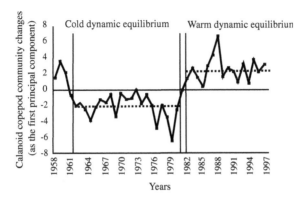

FIGURE 10.5. A shift from a "cold" to a "warm" Calanoid copepod community in the North Sea reflected by change in the first principal component of a multivariate data set. (From reference 9.)

over the last decades of the past century. For instance, strong biogeographical shifts have occurred in copepod assemblages, with a northward extension of more than 10 degrees latitude of warm-water species associated with a decrease in species typical of colder water.[10] This implies that the marked temporal change that was noted in the North Sea probably does in fact correspond to a spatial redistribution on a larger scale, just as Russell hypothesized earlier for the English Channel ecosystem dynamics. Still, this does not answer the question of whether alternative stable communities might exist along a temperature gradient. In any case, the sharpness of the changes observed indicates again that the biological communities seem to respond in a highly nonlinear way to change in the physical environment.

SARDINE–ANCHOVY CYCLES

Another much discussed example of marine ecosystem shifts is the remarkable alternation between sardines and anchovies (figure 10.7) along the coasts of Japan, California, and western South America and in the Benguela Current along the western coast of South Africa.[11,12] These shifts are especially relevant, as fisheries on small pelagic fish represent a very important contribution to the total world marine fish catches. Surprisingly, despite the dramatic nature of the shifts and

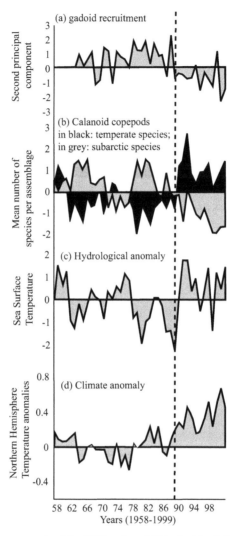

Figure 10.6. Some aspects of the 1987 regime shift in the North Sea. A drop in recruitment of cod and other gadoid fishes (a) was associated with a change from subarctic to temperate Calanoid copepod species (b) and to climatic indicators such as the sea surface temperature (c) and the northern hemisphere temperature anomalies (d). (From reference 9.)

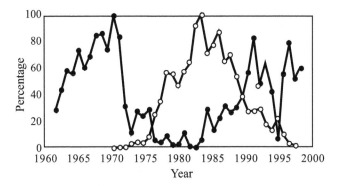

FIGURE 10.7. Catches of anchovy and sardine in the Humboldt Current ecosystem expressed as percentages of the maximum catch. Closed circles represent Northern and Central Peruvian anchovy, and open circles represent Peruvian and Northern Chilean sardine stocks. (From Alheit and Niquen; see reference 11.)

their economic importance, the responsible mechanisms remain rather unclear. Fisheries have certainly played some role in the collapses, but the fact that similar oscillations seem to have occurred long before substantial fisheries existed suggests that there is more to it. Indeed, circulation-induced changes between a "warm" and a "cold" mode seem to coincide with recent shifts.[12] However, these relatively smooth abiotic changes do not explain the tight coupling between sardine and anchovy abundances.

The fact that sardines and anchovies are never abundant at the same time is suggestive of competition (see the appendix, section A.4) where dominance by either sardines or by anchovies represents alternative stable states. It has been suggested that a particular kind of asymmetric competition known as the *school trap* might explain this pattern.[13] The argument goes as follows: Individuals of these pelagic fishes have a strong urge to join schools. It is known that when a species is abundant, it forms largely monospecific schools. In contrast, in periods when a species is rare, the individuals are found to join schools of other species. Now, if one supposes that the swimming speed and choice of foraging habitat of a school is tuned to the optimum conditions for the dominant species in a school, it follows that individuals of the species that is rare at that time and that are forced to join schools of the dominant species should have a reduced fitness.

Therefore, this school trap mechanism implies a kind of Allee effect (section 2.1), a positive feedback in rarity of either species. If this effect is strong enough, it may conceivably lead to alternative stable states. As always with positive feedbacks, even if the effect is too weak to cause true alternative stable states, it might well cause a threshold response around some critical set of conditions (section 14.3). In any case, such a feedback may explain how a gradual change in conditions (for example, a gradual change in plankton size and temperature) could lead to a sharp shift between dominant species. Importantly, changes in fishing pressure typically lag behind a change in availability of fish and will therefore tend to aggravate crashes of fisheries.

In conclusion, the sardine–anchovy cycles seem likely to be driven by physical forcing. However, the sharpness of the shifts suggests a nonlinear response to the physical environment that is generated by feedback loops in the community interactions and probably amplified by fishing pressure. Thinking about this on a more abstract level, it resembles the mechanisms behind glacial cycles where mild variation in radiation driven by orbital changes is amplified by positive feedbacks to produce sharp shifts in the system.

Collapse of Cod Populations

Cod populations represent another well-known example of how an economically important fish stock can crash (figure 10.8). Models suggest that the mechanism of exploitation can by itself cause the overexploited state to be an alternative stable state.[14] However, the lack of recovery decades after the closure of fisheries on Newfoundland cod has raised the question of whether other mechanisms may keep collapsed stocks from recovering. One possibility is that a sufficient adult stock is needed to control potential predators and competitors of their offspring.[15] Indeed, size-structured predator–prey interactions may well lead to Allee effects (section 2.1) that make it difficult to recover once the population has passed a critical threshold,[16] and other mechanisms might contribute as well to such Allee effects.[17]

It is interesting that a similar change has occurred in the scientific thinking about overgrazing in drylands. Here, it was long assumed

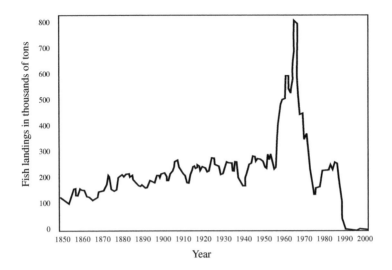

FIGURE 10.8. The collapse of Atlantic cod stocks off the east coast of Newfoundland in 1992 forced the closure of the fishery after hundreds of years of exploitation. (From reference 18.)

that overgrazing could cause a stable overexploited state.[19] However, just as Newfoundland cod did not recover upon the closure of fishery, some degraded lands did not recover upon removal of the animals. It was subsequently found that an additional positive feedback (between soil condition and vegetation presence) may keep the system irreversibly trapped in the degraded state.[20]

While the question of whether Allee effects may prevent cod from recovering is important, it is not the same as asking what has driven many cod populations to collapse. Surely, the role of fisheries is key. Archeological evidence suggests that cod has been an important fish for human consumption for thousands of years, and until the 1920s, it was not uncommon to find cod of up to 1.5 to 2 m in length, a size in accord with some old illustrations of drying cod the size of fishermen.[21] However, when mechanized trawling replaced the traditional hook-and-line fishery, the abundance and also the size of cod started to decline (figure 10.9). The preferential removal of larger individuals by fisheries is an important aspect, as only the larger individuals reproduce, and reproductive success of first-time spawners tends to be

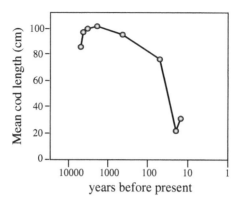

FIGURE 10.9. Decline in the mean length of cod caught from the coastal Gulf of Maine reconstructed from archeological evidence and fisheries data. (Redrawn from reference 31.)

poor.[22] Thus, a population can be pushed to a critical limit by removing the larger individuals.

On the other hand, fishery is not the only factor. As mentioned in the North Sea example, some studies suggest that successful cod recruitment depends on the availability of planktonic food of the right size at the right moment and may be favored by relatively low temperatures. In any case, cod is a species of cold waters, and increase in temperature may reasonably be expected to move the limits of its distribution toward higher latitudes.[23] The ability to withstand a high fishing pressure should therefore depend on the ocean climate. If fisheries push a population to a state in which availability of large-size individuals for spawning is greatly reduced, climatic conditions that are harsh for recruitment may well aggravate the situation. Therefore, it seems hardly surprising that the collapse of Nova Scotia cod at the Canadian coast has coincided with an increase in temperature.[24]

The danger in interpreting this relationship between climatic change and stock collapse is that it is easy to jump to the conclusion that climate determines almost everything when it comes to the state of marine ecosystems. First, it should be clear that fishery is pushing many stocks to the limit. In such a situation, climatic change may well trigger a collapse. One may then easily jump to the false conclusion that climatic change was *the* cause and that fishermen should be "let off the hook" when it comes to finding the driving factors.[25] However, a less severely exploited population with many old individuals will be much less affected by a bad recruitment year. Second, after a collapse is triggered, the newly achieved state often seems to persist, even if the physical conditions reverse. This was shown for the Pacific regime shifts (figure 10.7) but is also obvious in the dynamics of the Nova Scotia cod, where temperatures have reverted to normal without any sign of recovery of the cod population.[24] As argued, this is suggestive of a situation in which feedbacks in the system cause an Allee effect leading to an alternative stable state. Thus, physical forcing appears important but interacts with fishing pressure to drive the system that subsequently responds in a nonlinear way.

Last, it is interesting that cod collapse seems to have effects cascading all the way down the food chain.[26] This again goes against the notion that oceanic ecosystems are basically regulated in a bottom-up way and that circulation patterns basically drive the dynamics of productivity and ecosystem state. The tricky thing with bottom-up and top-down regulation is that they can both be strong at the same time and that their relative roles are not easily inferred from field patterns. Clearly, much of the variation in abundances we see in nature is bottom-up regulated, and marine systems are no exception. This is illustrated by the strong correlation between chlorophyll concentration and fish yields along the U.S. West Coast.[27] However, although this suggests that primary production largely determines what can be harvested on higher trophic levels, such empirical relationships cannot really tell us much about the importance of top-down forces. For instance, correlations between nutrient richness and the abundance of all trophic levels are commonly found in lakes.[28] Nonetheless, as discussed in section 7.1, top-down effects are strong in these ecosystems, to the point that lake managers have found that fish removal can boost large-bodied zooplankton, which then filters the water clear of excessive phytoplankton.

Given that many marine fish are being depleted, could top-down forces in the oceans be strong enough to imply similar cascading effects? Indeed, analysis of data from the Scotian Shelf[24] suggests that effects of the decline of cod and other large predators may cascade down the food web through small fish, crab and shrimp, zooplankton

Before   After

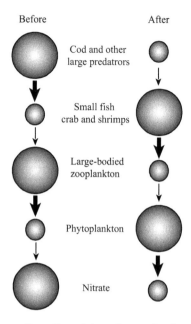

FIGURE 10.10. The cascading effect of the collapse of cod and other large predatory fishes on the Scotian Shelf ecosystem during the late 1980s and early 1990s.
The size of the spheres represents the relative abundance of the corresponding
trophic level. The arrows depict the inferred top-down effects. (From reference 26.)

and phytoplankton to nutrients (figure 10.10). One might suggest that
the observation of correlated changes can be a tricky basis for infer-
ring causal links. Indeed, a concomitant increase of the outflow of
low-salinity waters from the Arctic may have played a role in driving
change in the ecosystem too.[29] Nonetheless, a meta-analysis of cod–
shrimp studies has revealed that at least an increase in benthic invertebrates such as shrimp and crab has occurred almost everywhere that
cod stocks have collapsed on both sides of the Atlantic under different
climatic conditions.[30]

Overall, the results of the studies of causes and consequences of the
collapse of cod stocks are in line with the emerging view that marine
communities may be highly nonlinear systems. Such a view suggests
the need for a different look at management of marine ecosystems altogether. It implies that sharp irreversible change may sometimes re-

sult from gradually increasing fishing pressure and that the critical threshold for such change will vary with climatic conditions.

10.2 Coastal Ecosystems

Although the observations are suggestive, causality of regime shifts in the open ocean remains difficult to unravel. In contrast, coastal systems such as coral reefs, kelp forests, and sea grass beds are somewhat easier to study. Sharp changes have been documented in several of those systems, and evidence for regulating feedback loops and for an important role of fisheries and culling of marine mammals is relatively firm. In fact, there is little dispute on the importance of top-down effects here, and it has been shown that substantial impact of human exploitation of coastal populations with cascading ecosystem effects goes back even to prehistoric times. Recently, increased stresses such as eutrophication and habitat destruction have apparently set the stage for collapses of many coastal ecosystems, but the early loss of the regulating role of overharvested marine vertebrates may have tipped the balance in many cases.[31] Some coastal systems seem to recover relatively quickly and smoothly if exploitation pressure is released. For instance, overexploited shellfish communities of the productive Chilean rocky intertidal recovered to a state that seems not too different from the pristine condition in less than a decade after humans were excluded.[32] However, as you will see now, there are other systems in which human pressure has invoked collapse to a state from which recovery seems more difficult.

CORAL REEFS

Arguably the best known of all coastal ecosystems are coral reefs, with their fabulous biodiversity. Reefs worldwide suffer from a range of problems, including siltation, mechanical damage, bleaching events, and outbreaks of the coral-eating crown-of-thorn starfish. However, one of the most striking large-scale transitions is the shift of Caribbean coral reefs to a contrasting state covered by algae in the early 1980s. I briefly sketched what happened in section 1.1. Now I dig somewhat deeper into the question of how the regime shift may be explained. The picture emerging from a series of analyses over the past decades[33] is that (again) the cause is compound. Increased nutrient loading as a result of changed land use has promoted the growth rates of seaweeds (algae), but for a long time, herbivorous fish kept algal biomass under control. However, progressively intensive fishing reduced the numbers of large fish, and subsequently numbers of the smaller species, including important herbivores such as parrotfish and surgeonfish. Still, algal biomass remained low, as sea urchins took over the role of dominant herbivores. The urchins benefited from the reduced competition from fish and the high productivity of algae, and became very abundant. Eventually, the system collapsed when urchin populations plummeted because of a pathogen outbreak, resulting in less than 1% of the population surviving in most places. This meant that algae were released from the last remaining potent grazers, and the reefs rapidly became completely overgrown by macro-algae (figure 10.11).

There are several interesting aspects to the story of the collapse of the Caribbean reefs. First of all, it is a striking illustration of the *insurance effect* of biodiversity (section 5.3). When various species can fulfill a certain role, the system is less vulnerable than when a single species is left in a functional group, as was the case for the sea urchin, controlling the algae all by itself just before the collapse. A related aspect is the risk of disease. One could argue that the disease outbreak was a coincidence. However, as density of a particular species increases, the risk of an epidemic increases too. Thus, although there is always a large random component in the incidence of epidemics, one can imagine that they might be more likely in systems that are subject to a high nutrient load and also have a reduced biodiversity, simply because such systems tend to have high-density populations of key species, which promotes spread.

Last, it is remarkable that decades later, most reefs still have not recovered, even though urchins are around again. As in the case of the Newfoundland cod dynamics, this does not prove that the change represents a shift to an alternative stable state (see section 14.2). Nonetheless, it does look suggestive of a shift to an alternative stable state or, in any case, indicates that a return to the original state may be hampered by some mechanism. Potential candidates for such mecha-

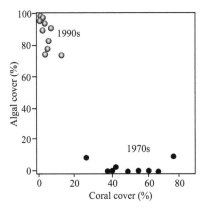

FIGURE 10.11. Censuses of the cover of algae and coral on Jamaican reefs illustrate the sharp contrast between the community in the years before the urchin collapse (1976–1980) and the situation after the transition (1990–1993). (From reference 34.)

nisms that stabilize the algae-dominated state are a low palatability of the large weeds (as opposed to the small new settlers that are grazed) and prevention of settlement of coral larvae by the thick algal cover. Theoretically, such feedbacks may well cause the algae-dominated situation to represent an alternative stable state. However, whether this could be the case depends on the strength of the feedback relative to other processes (section 14.3). Experiments seem a simple way to assess the stability of the alternative states, but it is tricky to extrapolate results from small-scale manipulations to the dynamics of entire groups of reefs. This is because the functioning of individual reefs actually depends on the regional situation, due, for instance, to the mobility of some of the key species and dispersion of coral larvae.[35] Also, reversal of a shift may require the presence of unexpected species that differ from the herbivores thought to be important in preventing a shift. For instance, on the Great Barrier Reef, it has been shown that exclusion of fish by means of experimental wire cages can allow a coral to become overgrown by seaweeds. Upon removal of the cages, only a particular batfish that swam in from the nearby ocean and that was previously regarded as an invertebrate feeder could remove the weeds again. The forty-three species of herbivorous fishes present in the local fauna played only a minor role.

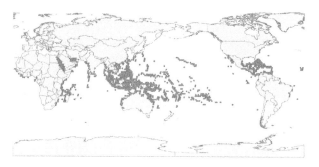

FIGURE 10.12. Distribution of coral reefs around the world. (Source: NASA.)

KELP FORESTS

The reef story has some remarkable parallels to what is believed to govern the dynamics of another important coastal ecosystem, kelp forests. Unlike coral reefs (figure 10.12), kelp forests are characteristic of the rocky shores of temperate waters (figure 10.13). They are impressive ecosystems, and their resemblance to forests is perhaps best illustrated by a quote from Darwin, who describes the kelp he encountered when he passed Tierra del Fuego in *The Voyage of the Beagle*:

> I can only compare these great aquatic forests . . . with the terrestrial ones in the intertropical regions. Yet if in any country a forest was destroyed, I do not believe nearly so many species of animals would perish as would here, from the destruction of the kelp. Amidst the leaves of this plant numerous species of fish live, which nowhere else could find food or shelter; with their destruction the many cormorants and other fishing birds, the otters, seals, and porpoise would soon perish also; and lastly, the Fuegian[s] . . . would . . . decrease in numbers and perhaps cease to exist.."

Indeed, the Fuegian people have gone extinct, although not because of destruction of the kelp. Nonetheless, deforestation of kelp forests has become a widespread phenomenon, and Darwin would surely have been interested in its causes and consequences.

Numerous studies have focused on the waxing and waning of kelp forests, but two reviews by Steneck and colleagues capture much of what is of interest in the context of this book.[37,38]

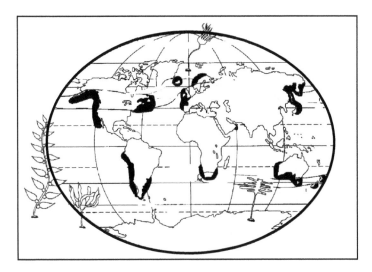

FIGURE 10.13. Kelp forests are distributed along the rocky shores of temperate water. (Modified from reference 36.)

A common theme across all studies of kelp deforestation is the role of sea urchins. These animals can completely wipe out kelp forests, transforming them to what is known as *urchin barrens*. Remarkably sharp boundaries are found between such barrens and remaining patches of forest.[39] This is suggestive of feedbacks stabilizing either state (section 14.1). Indeed, experiments have revealed that sea urchins can control kelp growth in the open areas, but migration of urchins into the kelp stands is prevented by foliage sweeping over the rocks in the border region. Thus, kelp stands seem to be able to defend themselves to some extent from their most important enemy by combining their flexible morphology with the energy of wave-generated surge.[39] Also, there is evidence that in some situations, crabs in the kelp stands may effectively prevent urchins from building up populations.[38]

Just as in coral reefs, algae-dominated patches can coexist with areas where the algae are controlled by grazers, but there can also be large-scale shifts that transform entire regions from a forested to a barren state or vice versa.[37] As an example, consider the history of kelp forests on the coast of Alaska (figure 10.14).[38] Kelp forests probably dominated this region since the last glacial period. However, when

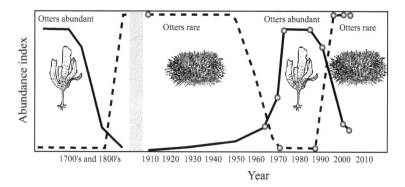

FIGURE 10.14. Shifts between a kelp-dominated state and a deforested state where kelp is controlled by sea urchins along the Alaskan coast, driven supposedly by changes in otter populations. (Modified from reference 37.)

hunting decimated populations of sea otters in the 1700s and 1800s, the kelp forests collapsed, as they were grazed away by sea urchins, now released from sea otter predation. Only in the twentieth century did kelp forests recover, as sea otters became legally protected. However, sea otters once again declined when killer whales shifted their diet to sea otters after the abundance of their preferred prey, seals and sea lions, had declined substantially.[40]

Another example of sequential large changes in kelp forests comes from the Gulf of Maine.[38] Here, the pristine state that persisted for at least 4,000 years is one dominated by lush kelp forests. In this state, sea urchins were controlled by abundant large predatory fish such as cod and haddock. As fisheries decimated these apex predators, kelp forests were progressively grazed down by sea urchins. In the 1980s, only a few relicts of the once extensive forests remained in urchin-free shallow and turbulent zones. However, human exploitation of marine resources would once again change the fate of the forests in an unexpected way. In 1987, fishing of the green sea urchin for the Japanese sushi market began and quickly depleted populations from vast areas along the coast. When the populations dropped below an apparent threshold biomass, their grazing pressure could no longer control macro-algal recruitment, and phase shifts back to a kelp-dominated state occurred. Interestingly, the new kelp forests appear to persist even in parts were the urchin fishery has become prohibited. Evidence

suggests that in the absence of cod and other former apex predators, crabs may now control urchins. Further north, along the coast of Nova Scotia, a fascinating variant to the Gulf of Maine dynamics has been observed. Here, urchins and kelp slipped into an oscillating state with periodic urchin-disease outbreaks that make the system flip between a forested and a barren state.[37,41]

In summary, kelp forests and overgrazed barrens seem to represent alternative stable states. Although both states often coexist in space, with sharp boundaries between them, large-scale shifts have occurred that swept entire regions from a predominantly forested to a largely barren state or vice versa. All those shifts appear related to human-induced changes in predation pressure on sea urchins, either through cascading effects of harvesting urchin predators such as sea otters of predatory fish, or through direct harvest of urchins.

OYSTERS IN ESTUARIES

Estuaries represent the tidal mouths of rivers. They have increasingly become loaded with silt, nutrients, and pollutants in most parts of the world. Moreover, the adjacent land tends to be densely populated, and exploitation of shellfish and fish in the region is often intense. No wonder that estuaries are considered to be among the most degraded of all marine ecosystems. From the viewpoint of ecosystem stability, it is interesting that there can be positive feedbacks between biota and the physical conditions that could lead to runaway change in some situations.

First let us look at the history of Chesapeake Bay on the U.S. East Coast (figure 10.15).[31,42] In the mid-eighteenth century, when widespread land clearance for agriculture by European colonists began, increased sedimentation and input of organic carbon started to affect the ecosystem. Gradually, plants and algae growing on the sediments became less abundant, and phytoplankton became more abundant. However, the vast oyster reefs present filtered the water intensively. This changed when mechanical harvesting with dredges started to replace the small-scale artisanal exploitation that had gone on for millennia without doing much damage to the reefs. Dredging ruined the reefs, and catches dropped to a few percent of the peak values by the

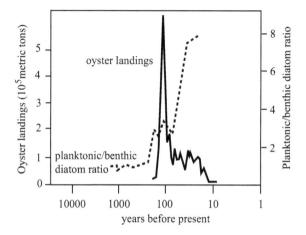

Figure 10.15. Nutrient loading, but also mechanized trawling of filter-feeding oysters, has caused a shift of the Chesapeake Bay estuary from a state with relatively clear water and predominantly benthic plants and algae to a state with phytoplankton dominance and widespread anoxia. (Redrawn from reference 31.)

beginning of the twentieth century. With the collapse of the oyster fishery, the ecosystem changed profoundly. In the absence of those filter feeders, phytoplankton biomass became very high, and as in other aquatic systems, decomposition of the large amount of organic material started to lead to anoxia. Also, oyster diseases became common in the new situation. Recovery of the oyster populations now appears to be prevented by these factors. For instance, experiments in Pamlico Sound, a similar estuary, showed that oysters grow well and are not affected by disease if they are elevated above the zone of deeper water where hypoxia occurs in summer.[43] Since dense populations of oysters and other bivalves can graze phytoplankton so efficiently that they limit blooms, thus preventing symptoms of eutrophication including hypoxia, this suggests a positive feedback through which filter feeders may improve conditions for their own survival. Nutrient loading can push such systems to an increasingly eutrophic state, but it seems that oyster beds have been able to safeguard their own persistence for a while through *top-down control of water quality*. Massive harvesting of the oyster beds appears to have triggered the shift to an alternative stable state in which return of the oysters is prevented by hypoxia.

Sea Grass Fields

Extensive sea grass fields have long been an important feature of many shallow coastal systems. In tropical and subtropical bays, in lagoons, and on continental shelves, vast meadows once provided forage and habitat for large mammals such as dugongs and manatees and for sea turtles. Also, they harbored diverse communities of fishes and invertebrates. Mass mortality of sea grass has changed many of these ecosystems drastically. As in shallow lakes, eutrophication has been identified as an important causal factor. However, the sea grass story is special in some important aspects.[31,44] One marked difference is that in some places, anoxia may lead to massive sudden die-off in sea grass. This implies a parallel to the oyster case. Anoxia is more likely to occur if organic matter production increases and if water circulation is poor. Once die-off starts, it appears to be a self-propagating process, as decomposition of the dead material takes up even more oxygen. Excessive nutrient input from neighboring rivers tends to promote productivity and organic matter production, and therefore the risk of anoxic events. However, the formerly abundant herbivores may have helped to prevent such a situation by keeping the biomass low and creating a more open structure that allows water circulation. For instance, herds of dugongs comprising tens of thousands of individuals were observed around 1870 at Wide Bay, in tropical Australia. Colonial exploitation of these huge mammals for flesh and oil left only a fraction of the original populations. Also green turtles have supposedly closely cropped the meadows of what was also known as "turtle grass" in those days. It is therefore conceivable that circulation was better before natural consumer populations became repressed by human exploitation (figure 10.16). In summary, anoxic die-off of sea grass fields in shallow warm waters seems to be a threshold phenomenon, because of a positive feedback once the die-off starts. Such a periodic collapse caused by the building up of an organic matter pool is in a sense analogous to the internal eutrophication that may cause cyclic collapse of the vegetated state in some lakes (section 7.2).

Sea grass fields have also entirely disappeared in many of the temperate waters where they were once abundant, such as the tidal flats of The Netherlands. As elsewhere, eutrophication and diseases seem important factors, but a clear picture of the causality is still lacking.

FIGURE 10.16. Green turtles on a wharf in Key West. Vast populations of these animals once cropped sea grass in the Americas and Australia. Their elimination is thought to have reduced the resilience of sea grass meadows against autocatalytic anoxic die-off. (From reference 45.)

Especially puzzling is the question of why the once extensive fields have not come back, even though water quality has improved. The reasons are basically unknown. One explanation that has been suggested is the erosive force of the water. One might speculate that extensive dense fields reduce turbulence enough to protect the plants, whereas it may be difficult for a small colonizing patch to avoid being damaged and washed away by currents and wave action. Also, much as in shallow lakes, the plants may promote water clarity, implying a positive feedback on growth through the underwater light climate.[46]

Erosion of Benthic Communities

Erosion is a mechanism that may explain Allee effects in the colonization of barren substrates by plants in terrestrial and aquatic ecosystems. Sea grass fields are one possible example out of a more general category of aquatic systems in which the erosive force of waves and currents can make the first steps in colonization difficult. This includes the sediments of shallow freshwater lakes that may be stabilized by a microbial community and subsequently by submerged vegetation[47] and also various marine examples.

First, let us consider a shift in the bottom fauna of a part in the North Sea known as the Friesian Front.[48,49] This area is located on the Dutch Continental Shelf at the transition between the permanently mixed waters of the Southern Bight and the summer-stratified water masses of a deeper part known as the Oyster Grounds. Owing to the local deposition of particulate matter and settling fresh algal detritus, this area is rich in bottom fauna.[50] The brittle star *Amphiura filiformis*, used to be extremely abundant in the southern part of the Friesian Front. In the period between 1984 and 1992, its densities ranged from 1,433 to 1,750 individuals per square meter.[51] However, in the years after 1992, the population collapsed to a small fraction of the original densities (figure 10.17). Associated with this sharp shift, the entire macrobenthic fauna community changed, and the burrowing mud shrimp *Callianassa subterranea* became a dominant species.[49]

In the search for a possible explanation of such a transition, as always, one has to realize that a sudden shift in a community is not necessarily related to alternative stable states (section 14.1). An obvious alternative explanation is that there has been a shift in an important controlling factor (for example, temperature). Alternatively, the system may have been hit by a severe disturbance from which it is simply slow to recover. There is no evidence that external control factors such as the algal biomass of the North Sea or the state of the North Atlantic Oscillation (NAO) are systematically different in the new state from what was observed during the brittle star state.[49] Also, the local brittle star collapse does not correspond to known large-scale regime shifts in the North Sea around 1979 and in the 1980s that have been linked to changes in the abiotic conditions.[9] The system may well have been hit by a disturbance such as a trawling activity or an extreme shear stress due to storm and tidal current action. However, in view of the generation time and growth rates of these brittle stars, the years since the shift should have offered sufficient time for the brittle star populations to recover. The persistence of the new community state therefore suggests that there is a stabilizing mechanism that maintains the

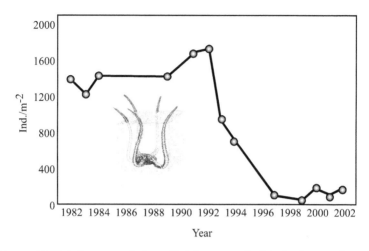

Figure 10.17. A dense population of brittle stars at the Friesian Front collapsed in the early 1990s. (Redrawn from reference 49.)

shrimp-dominated situation and prevents the brittle stars from re-covering to the original densities.

Field measurements reveal high concentrations of suspended sed-iments in the Friesian Front region. This indicates that sediment ero-sion and resuspension are important features of the new ecosystem state. The intuition of the researchers was that the dense "fields" of brittle stars might have stabilized the sediment. Indeed, they could demonstrate experimentally that sediments inhabited by brittle stars were less susceptible to erosion by tidal currents and wave forces than sediments inhabited by shrimp.[52] The burrowing shrimp apparently thrive under the turbulent conditions of the new state, but settlement of the fragile juvenile brittle stars seems unlikely on such unstable and frequently resuspended sediments. A model confirms that a positive feedback between brittle star abundance and sediment stability may plausibly lead to alternative stable states.[48] It also shows that, as always in such a situation, a stochastic event such as a storm may trigger a shift between the states (figure 10.18).

Another example of a shift between alternative states in marine benthos comes from tidal mudflats. Here benthic diatoms can stabi-lize the surface, preventing erosion. Once stabilized, the mudflat sur-

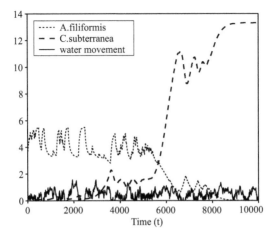

FIGURE 10.18. Simulated state shift in the benthic community at the Friesian Front from dominance by brittle stars (*A. filiformis*) to dominance by burying shrimps (*C. subterranea*). Natural variability in the shear stress caused by water movement triggered the shift to the new state, in which frequent erosion is assumed to prevent resettlement of brittle stars that could stabilize the sediment once they are sufficiently abundant. (From reference 48.)

face promotes better diatom growth. It has been demonstrated by a combination of experiments and models that the resulting positive feedback between diatoms and sediment characteristics can lead to alternative stable states: erosion-prone barren versus diatom covered.[53] Of course, it is easier to test a hypothesis in an easily manipulated system such as this that can even be brought to the lab, without losing too much of its realism, than in most large-scale systems. For instance, a field experiment to test the theory on the Friesian Front shift is impossible, in view of the rough seas in the area.

10.3 Synthesis

In summary, marked shifts are common over a range of marine ecosystems, but their character differs widely between cases, and mechanisms are often quite poorly understood. Regime shifts documented in the open ocean and on the high seas seem mostly caused by extremes

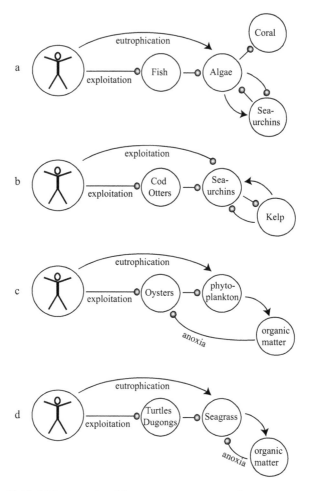

FIGURE 10.19. Main impacts of humans on some coastal marine systems and some ecological feedbacks that may be responsible for nonlinear responses or alternative stable states. (a) Coral reefs: The double feedback from macro-algae to sea urchins denotes the possibility that while algae are profitable food for sea urchins, this may be less so if the algae have become big and unpalatable. (b) Kelp forests: The double feedback is because fully grown kelp may keep out urchins by a mechanical whiplash effect and by harboring urchin predators such as crabs. The mass-exploitation of urchins by humans is relatively recent and has lead to recovery of kelp forests in some places. (c) Estuarine oyster banks: By keeping phytoplankton biomass low, oysters may help prevent anoxia that would otherwise wipe out their population. (d) Sea grass: Buildup of organic matter may cause autocatalytic mass mortality beyond a threshold.

within the natural variability of the physical environment. However, there is evidence that the biological system responds to such forcing in a nonlinear way. It appears that feedbacks in the biological communities are strong enough to cause threshold-like responses to changing conditions. Although true alternative stable states may exist, this is difficult to prove. In addition to the physical climate, fisheries affect many marine systems strongly. It follows from the basic theory (section 2.1) that thresholds for fishing pressure and physical conditions must be mutually dependent. Thus, safety limits to exploitation depend on the physical regime, and climate sensitivity depends on fishing pressure.

In coastal systems, sharp shifts are also commonly observed, and in several cases, mechanisms have been better unraveled than in the open ocean. Exploitation of key species and eutrophication are common drivers of change in coastal systems (figure 10.19). Again, critical limits of different factors are likely to be mutually dependent. Unlike in the open ocean, alternative states may have a local character in coastal systems in the sense that patches in different states can coexist side by side. Nonetheless, large-scale transitions occur too. This can be related to domino effects or to positive feedbacks between the local biota and the large scale "mixed part" of the system (the water, fish, plankton, and so on).

CHAPTER 11
Terrestrial Ecosystems

Although terrestrial ecosystems have traditionally been studied in much more detail than most marine or freshwater systems, the issue of stability shifts is less prominent in the associated ecological literature. Nonetheless, good examples of critical transitions have been documented. As you will see, there is a division into roughly two classes, much as in the marine realm. There are small-scale patterns and shifts, but there is also a possibility for sharp regional transitions. The latter have to do with a feedback between the vegetation and the regional climate. I will highlight examples of both, first from warm and dry regions of the world and then from colder systems. In addition, I briefly discuss two broader phenomena that correspond to critical transitions: the collapse of species as slow ongoing fragmentation of their habitat exceeds a critical threshold and the outbreak of epidemics when a critical fraction of susceptible individuals is exceeded.

11.1 Vegetation—Climate Shifts in Dry Regions

Climate is the main factor determining the natural distribution of ecosystem types (*biomes*) over the surface of the Earth. Temperature determines a lot, as you will see in some of the next sections, but humidity is another major issue. In much of the tropics and subtropics, precipitation determines to a large extent where we find deserts, grasslands, savannas, or forests. Interestingly, vegetation itself can also promote precipitation in some regions. This implies a feedback that may in some situations be strong enough to lead to alternative stable states, implying the possibility of large-scale transitions once a critical threshold is passed. With climatologist Victor Brovkin and vegetation ecologist Milena Holmgren, I reviewed this phenomenon, and what follows is based on our common work.[1]

THE MECHANISMS

The view that vegetation cover may affect the regional climate became known particularly with the work of Jule Charney.[2] He argued that the high albedo of unvegetated dry land induces a sinking motion of dry air and therefore suppresses rainfall. Low precipitation, in turn, results in little vegetation cover. This feedback may thus maintain a desert that is self-sustaining. On the other hand, once there is more vegetation, the feedback works the other way round. Since vegetation is darker than sand, it absorbs more heat, and the temperature gradient between land and ocean increases. This amplifies upward air motion over the desert and can promote monsoon circulation, bringing moist air from sea to land. As a result, the rainfall in a region may increase. Later studies have elaborated on this insight. For instance, it was shown that in the Sahel–Sahara region, the distribution of monsoon rains is influenced by vegetation, especially by the West African forests.[3,4]

Another way in which vegetation may influence the local climate is through transpiration. Especially trees are important,[5] as some have roots long enough to reach deep groundwater that would otherwise be unavailable for plants. A large part of this deep groundwater taken up by trees can become available to other plants as it falls down again locally as rain. The relatively dark canopies absorb much radiation. This energy is largely used for evaporation of water. The rising water vapor releases heat if it condenses in cumulus clouds, leading to further air lift, more clouds, and eventually possible showers. As a result of this convective mechanism, the local climate may be much cloudier and rainier than if forests would be absent. Conversely, tropical deforestation may result in a decrease in such moisture recycling.[6]

Since water is a major limiting factor for vegetation in many regions of the world, a stimulating effect of vegetation on precipitation implies a positive feedback. Just as for the particular mechanism suggested by Charney, one can easily imagine that this could cause a vegetated wet state and a barren dry state to be both stable and self-reinforcing. However, whether this really happens depends on the strength of the feedback and on the heterogeneity of the landscape (see chapter 5 for a basic discussion of the effect of heterogeneity). A graphical model of the feedback between terrestrial vegetation and regional climate may illustrate this.[1,7,8] The model has two essential components: the response of vegetation cover to precipitation and the response of precipitation to changes in vegetation cover. To start with the first component, let us assume that a given microsite will have vegetation whenever precipitation exceeds a certain critical threshold. In most landscapes, this critical precipitation level will vary between microsites as the sites differ in aspects such as soil fertility and also in actual microsite moisture level depending on their topographic position (for example, valley versus hill slope or north- versus south-facing slopes). Thus, whereas completely homogeneous landscapes would theoretically shift from barren to vegetated at a single critical precipitation level, heterogeneity of a landscape should result in a smoother increase of vegetation cover over a certain range of precipitation levels, as some microsites become suitable at lower precipitation levels than other sites. In summary, we should expect vegetation to increase with precipitation sharply around a critical precipitation in homogeneous landscapes, but the response would change into a smooth sigmoidal function in more heterogeneous landscapes (the line $V' = 0$ in figure 11.1). With respect to the second component (response of vegetation to vegetation), let us assume as a first approximation that an increase in vegetation cover produces a linear increase in regional precipitation (the line $P' = 0$ in figure 11.1), which for regions such as the Sahel and the Amazon is roughly what is suggested by more elaborate simulation models.[7,8]

If we plot the hypothesized response of vegetation cover to precipitation and of precipitation to vegetation cover together, we have a simple graphical model of the dynamic interaction between vegetation and climate (figure 11.1). Intersections of the two response lines represent equilibria of the interactive system. It can be seen that multiple intersections, and hence multiple equilibria, can arise only if

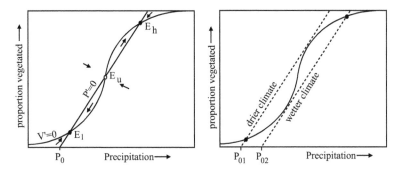

FIGURE 11.1. If vegetation enhances the amount of precipitation compared to the level in absence of vegetation (P_0), the resulting tilted equilibrium precipitation ($P' = 0$) line may intersect with the vegetation equilibrium line ($V' = 0$) at three points (panel a). The arrows indicate the direction of change if the system is not in equilibrium and show that the intermediate equilibrium point (E_u) is an unstable saddle point, whereas the low and high biomass points (E_l and E_h) are stable. If general climatic conditions are sufficiently dry or wet, situations with single stable intersections may arise (panel b). Note that for simplicity, it is assumed that the slope of curve P' remains unchanged, which is not necessarily the case. (From references 1 and 14.)

vegetation cover rises steeply enough around a certain critical precipitation level and if vegetation has a sufficiently strong effect on precipitation (implying that the $P' = 0$ line is not too vertical).

Although this simple graphical model illustrates the basic idea, strong cases for the existence of alternative attractors can be made only on the basis of more detailed analysis of regional climate systems, using data on their recorded dynamics, quantitative information on the main processes involved, and realistic simulation models that pull all the information together.

THE SAHEL–SAHARA

A well-studied example of a region where positive feedback between precipitation and vegetation is considered important is the Sahel–Sahara.[4,7,9,10] As mentioned briefly in the introductory chapter (section 1.2), paleo-reconstruction for this region has provided evidence of shifts between contrasting vegetation–climate states in prehistoric times. During the early Holocene, much of the Sahara was far wetter than today, with extensive vegetation cover, lakes, and wetlands.[11] Marine-core data indicate that around 5,000 years before the present, an abrupt switch to a desertlike state occurred in western North Africa.[12] Most probably, this change has been caused by a gradual reduction in insolation resulting from subtle variation in the Earth's orbit. This has led to some decrease in the temperature contrast between ocean and land, which weakened the monsoon circulation, resulting in a reduction in precipitation. While some computations suggest that the feedback may cause a sharp response but no true hysteresis,[13] other simulations[7,10] indicate that the feedback effects of vegetation on this monsoon rain system may lead to true alternative stable states. This could thus explain why the regional vegetation–climate system apparently stayed rather constant until a rapid transition to the alternative state occurred. Put simply, vegetation cover helped maintain the monsoon circulation despite the reduced insolation for some time. However, once a threshold was reached, drier conditions and loss of vegetation started promoting each other in a runaway process, driving a critical transition toward a desert state.

THE AMAZON

Another example of a situation in which hysteresis may result from a feedback between vegetation and the regional climate is the Amazon.[8,14] Sediment analyses indicate that an open vegetation and closed forest have both existed over long time spans, whereas intermediate states have been rare. As in the Sahara case, model analysis indicates that a feedback of vegetation on precipitation may plausibly cause alternative equilibria that would explain this pattern. Because the transpiration flux is greater for forest than for savanna, presence of forest amplifies local moisture recycling and may lead to an increase in precipitation during the dry season. Depending on the precise relationships, this feedback can indeed cause forest and savanna to represent alternative stable states. A notable example of a perturbation that could induce a large-scale stability shift would be deforestation. Beyond a critical limit, this could bring the system into the attraction basin of the alternative equilibrium, causing a regional collapse from a wet forested state into a dry savanna. In the new, drier state, eastern Amazonian forests would be replaced by savannas, a semidesert area

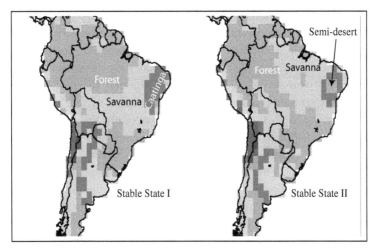

FIGURE 11.2. Model analyses suggest that under the current overall climatic conditions, the distribution of biomes in the Amazon region can stabilize into two alternative states depending on whether they start from a barren or from an overall vegetated state. (From reference 14.)

would appear in the driest portion of northeast Brazil, and the Atlantic tropical forest would expand northward in subtropical South America (figure 11.2). Ongoing deforestation in the region might thus pose a serious threat of critical transition toward this drier state.

11.2 Small-Scale Transitions in Semiarid Vegetation

Although feedback of vegetation on the regional climate may be a key mechanism behind dramatic nonlinear responses of the vegetation–climate systems, on a smaller scale, there may be alternative stable states in vegetation as well. Indeed, this seems almost the rule rather than the exception in vegetation of dry regions of the world.

INERTIA AND SHIFTS IN SEMIARID VEGETATION

Persistence of the spatial distribution of different vegetation types on seemingly homogeneous landscapes is one of the observations suggesting that the contrasting vegetation types represent stable states.

For example, comparison of the extension of woody patches in an otherwise herbaceous vegetation in semiarid Chile revealed virtually no change over a 30-year time span,[15] indicating that both types of vegetation are highly resilient. Despite considerable variation in weather over the 30-year period, all woody spots remained woody and none of the open herbaceous areas was invaded by shrubs or trees. Occasionally, however, such status quo between woodland and open areas in dry regions is punctuated by remarkable change.

Herbivore mortality events are one of the factors that may trigger forest expansion. The reconstructed long-term history of African savanna dynamics provides a good example.[16] A massive rinderpest epidemic that swept through the region in the 1890s greatly reduced numbers of ungulates, allowing large-scale woodland expansions. The resulting extensive woodlands were stable. They dwindled again only 30 to 50 years later, when they were eliminated by human-induced fires. Subsequently, the open landscape was maintained by renewed top-down control by recovered populations of elephants and ungulates. Similarly, devastation of the rabbit population by myxomatosis has led to massive tree recruitment in many places. A well-documented example is the shift of the open landscape of Silwood Park, London, to a dense oak woodland following the myxomatosis epidemic of 1955.[17] Although the rabbit population has recovered, the tree stands persist because they have achieved an escape through their size from herbivory by rabbits. Rare periods of extreme rainfall may also trigger woodland expansion. This is especially obvious in semiarid regions where ENSO events can lead to increased rainfall of an order of magnitude or more. Tree ring analyses in Peru have revealed that indeed rare, wet El Niño periods have been key years for tree recruitment,[18] while aerial photographs and satellite images confirm that massive woodland expansion can occur following extreme rainy El Niño events.[19]

If such forests need repeated extreme events to allow rejuvenation, one cannot consider them a "true" alternative stable state. However, there appear to be situations in which such newly established vegetation maintains itself in the long run even if no new extreme rainy periods occur. What could be the mechanisms involved in stabilizing the vegetated versus the barren state? Positive effects of plants on their own microsite growing conditions are thought to play a major role in determining the stability of plant cover in dry regions.[20,21] Plants and their deposited dead organic material allow precipitation to be absorbed by the topsoil and become available for uptake by plants. When vegetation cover is lost, infiltration decreases, and rainwater is lost through runoff and evaporation. Another reason why persistence of existing vegetation requires less precipitation than vegetation expansion is the drought sensitivity of seedlings. Even when plant recruitment does not occur in open spaces, seedling establishment is often found to be possible under the shade of nurse shrubs or trees, allowing rejuvenation and long-term persistence of an existing vegetation. One of the reasons for this nursing effect is an improvement in the seedling water conditions.[22] An extreme example is the situation in coastal areas, where trees can act as moisture traps and where fog from the ocean condensates, resulting in water dripping down to the ground and enhancing the microclimatic moisture.[23] However, a more subtle effect of plant canopies on water conditions is important in a wide range of dry ecosystems. In the shade of a tree or other nurse plant, air and soil temperatures are lower, and water content of the superficial soil layers tends to remain higher.[24] Therefore, seedlings experience lower thermal and water stress.[25] Other factors often contribute to this nursing effect—for example, soil nutrient levels can be higher and herbivory levels lower under the nurse plant.[26] The nursing effect, together with the fact that adult shrubs are often less sensitive to drought and herbivory explains why mature woody vegetation may sometimes persist and rejuvenate in climatic regions where establishment of seedlings in the absence of nursing shade of woody vegetation is impossible.

Much like the feedback on precipitation, the positive effects of plants on microclimatic moisture conditions may lead to alternative stable states.[20,27] A graphical model illustrating how this can work is shown in section 2.2.

SELF-ORGANIZED VEGETATION PATTERNS

A quite different mechanism that may lead to alternative attractors is driven by transport of nutrients and water from the barren land to the vegetation patches. Models suggest that this may lead to a situation where self-enriching vegetation patches represent an alternative stable state to a completely desertic condition devoid of perennial vegetation.[28] The mechanism is intuitively straightforward. The vegetated islands of fertility concentrate water and nutrients to a level that allows local vegetation persistence even though the average water and nutrient levels in the landscape are too low to support vegetation. Once vegetation patches are lost, the concentration mechanism is gone too. Consequently, recolonization of the barren landscape is possible only if precipitation levels have become high enough to support vegetation on average landscape conditions (explained in more detail in section 4.1). There are now numerous examples of characteristic vegetation patterns in arid areas (figure 11.3) that match the self-organized patterns predicted by the spatially explicit models remarkably well.[28] This makes the theory of alternative attractors and hysteresis in these systems quite plausible.

Alpine Tree Lines and Lowland Tree Islands

Last, we may ponder the question of why natural forest boundaries are often so distinct. As a general rule, sharp boundaries may be indicative of the existence of alternative stable states feedback (section 14.1), and indeed, sharp forest boundaries have been explained from positive feedback in several studies. For instance, alpine tropical tree lines can result from photo-inhibition of seedlings.[29] Adult trees protect the young plants against excessive radiation, which is a problem increasing at high altitudes because of low temperatures. Thus, rejuvenation is promoted in the forest, securing long-term stability, but invasion of the open area is prohibited.

Sharp tree lines are also observed in lowlands, where they may have a quite different explanation. Extensive flat wetlands such as the Florida Everglades and the Brazilian Pantanal are often dotted with *tree islands* (figure 11.4). The soil in such islands is enriched and elevated compared to the surrounding area, which reduces flooding incidence and therefore facilitates tree growth. However, on the tree islands, termite activity and other processes also promote accumulation of soil material, implying a positive feedback.[30]

In summary, sharp boundaries, persistence of patterns, and incidental shifts to a contrasting state seem common in semiarid as well

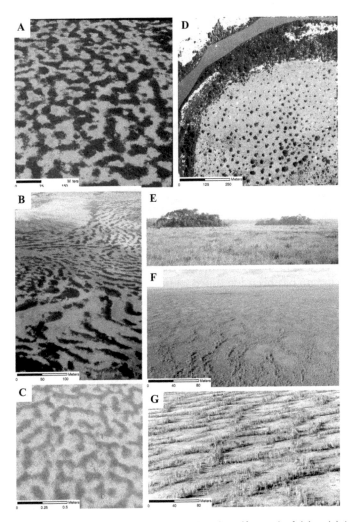

FIGURE 11.3. Vegetation patterns that are likely to be *self-organized*. (a) to (c) Arid ecosystems: (a) Labyrinth of bushy vegetation in Niger; (b) striped pattern of bushy vegetation in Niger; and (c) labyrinth of a perennial grass in Israel. (d) and (e) Savanna ecosystems: Aerial and ground photographs of spots of tree patches in Ivory Coast and French Guiana, respectively. (f) and (g) Peatlands: Regular maze patterns of shrubs and trees in western Siberia. (From Rietkerk et al.; see reference 28.)

as some other tropical and subtropical regions. Mechanisms may differ from case to case, but positive feedbacks between vegetation and the microclimatic or soil conditions have been identified in many cases. So far, the work on such small-scale feedbacks and shifts in dry regions has been poorly connected to the larger-scale feedbacks between vegetation and climate that may lead to shifts on regional scales discussed in the previous section. New insights may emerge as those research fields are further integrated.[1]

11.3 Boreal Forests and Tundra

While drought is a major problem for vegetation over much of the globe, conditions become harsh for different reasons as we approach the poles. As in dry regions, a positive feedback between vegetation and climate has been identified in those cold areas. After looking at this climatic feedback, we ponder the question of how boreal forests may irreversibly shift to lichen fields and look into the famous outbreaks of spruce budworm.

Forest–Climate Feedback in the Boreal Region

The extensive boreal forests in Canada, Scandinavia, Russia, and other places eventually give way to an open landscape as one comes closer to the pole and temperatures drop too deeply. In the open tundra planes, conditions are too harsh for trees to grow. Historically, it was thought that the extent of the boreal forest was limited by climate here. However, it has become clear that this is only part of the story. The forest itself also affects the climate, as it reduces the albedo in those snowy landscapes. This leads to a positive effect on local temperatures, earlier snowmelt, and a longer growing season. The effect may be strong enough to lead to local alternative stable states. Thus, in some regions, deforestation might lead to cooling that is sufficient to prevent regrowth of trees in those parts.[31] The other side of the coin, of course, is that warming of the region with global climate change might be amplified if it promotes forest expansion. Model analyses suggest that on a larger geographical scale, we should not ex-

FIGURE 11.4. Tree islands in the Brazilian Pantanal appear to represent self-reinforcing alternative states to the surrounding herbaceous vegetation matrix. (Photo by the author.)

pect a runaway process that is strong enough to lead to alternative stable states caused by this feedback, but indicate that it is nonetheless a significant regional amplifier of global warming.[32] Clearly, if one wants to estimate such a regional climatic feedback, it is important to put it into the context of other regional processes (figure 11.5). Very important in the Arctic is the feedback involved in the loss of Arctic sea ice. Open water absorbs much more heat than snow. Therefore, loss of sea ice too implies an accelerated warming of the region. On the other hand, there are negative feedbacks that tend to dampen the acceleration of warming. For instance, during the growing season, when trees have a denser, more productive canopy than herbaceous plants and moss, they transpire more water, implying a cooling of surface air in forests as compared to tundra. Also, there is a negative feedback in the northern ocean, as increased sea surface temperatures imply less dense water. Just like increased influx of freshwater, this could slow down the thermo-haline circulation (section 8.3), implying less heat transport toward the region and therefore eventually colder sea surface temperatures. If one uses reasonable models to integrate all those effects, it can be seen that the positive feedbacks including the tree–climate feedback prevail,[32] even if they do not lead to critical transitions between alternative stable states.

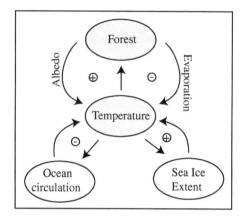

FIGURE 11.5. Forest in the boreal zone leads to higher temperatures that further promote forest expansion. However, this (shaded) feedback is part of a larger set of feedback loops in the regional climate system. Overall, the positive feedback predominates, implying among other things a regional amplification of global warming but no large-scale alternative stable states.

Such linked feedback systems involving a regional interaction between vegetation on climate occur in other parts of the globe too. A perhaps counterintuitive consequence is that terrestrial vegetation can affect ocean circulation patterns. Simulation studies suggest that this is a substantial effect that may be important in explaining some of the drastic Earth system transitions in the past.[32]

LICHENS VERSUS TREES

Looking in more detail, the boreal forest is not the homogeneous ecosystem it may appear to be at first glance. For instance, in Canada between the closed forest in the south and the tundra in the north, there is a distinct zone with an open forest known as *lichen woodland*. In this lichen zone, trees are sparsely distributed and the ground is covered by a mat of lichens dotted with some small shrubs. In Quebec in the Parc des Grand-Jardins (Park of the Big Gardens), such lichen systems occur in remarkable large patches, 500 kilometers south of the limit of their usual range. The history of these patches has been reconstructed using cores from scattered peats.[33] Plant fossils, charcoal layers, remains of insects, and pollen from those cores tell a fascinating story. The lichen patches appear to have a long history. All of them were once closed forest and came into existence suddenly. Their inception dates lie between 580 and 1440 years ago, and once in place, they never switched back to the forest state. This remarkable persistence, together with the fact that they occur in the same climatic and soil conditions as the adjacent closed forest suggests that the lichen lands represent an alternative stable state. It appears that a closed lichen mat prevents tree recruitment almost completely, explaining its stability in the otherwise forest-dominated region.

On the other hand, closed forest remains the dominant vegetation type in this part of the country, and such forest is apparently not easily taken over by the lichens. So what then may explain the inception of those large lichen lands in the Parc des Grand-Jardins centuries ago? A closer look at the peat-cores reveals that all of the shifts from forest to lichens coincide with abundant remains of head capsules of spruce budworm, an insect famous for incidental massive defoliation of the spruce forests. However, such outbreaks have always occurred in the entire zone, whereas the lichen lands have arisen only in a particular area and in a particular period. Thus, insect outbreaks cannot be the whole explanation of shifts. Charcoal layers suggest that fire may be the other part of the story. Lichen lands are found only in a zone characterized by higher fire incidence, and the cores show that the transformation from forest to lichen requires a coincidence of an insect outbreak, followed by an intense fire. Thus, the inception of the lichen lands in these boreal forests fits in the list of ecological surprises where ecosystems are tipped into an alternative state by a rare compound perturbation.[34]

INSECT OUTBREAKS

The insect outbreaks in the boreal forest merit a closer look too. As discussed later, disease outbreaks are excellent examples of critical transitions where a runaway process takes over once a critical threshold is passed, and the budworms are no exception. They are the larvae of a moth, and they specialize at eating the new green needles of conifer trees. Spruce budworm outbreaks have large implications for the forest industry. Therefore, these insects have quite a long history of being subject to management efforts and studies. The outbreaks happen irregularly, with a time varying from 30 and 100 years between events. The insect destroys large areas of mature softwood forests, principally spruce and fir. As long as the forest is in an earlier phase, foliage is relative open. This makes birds and other predators efficient in finding and controlling the budworms. However, as the foliage becomes denser, food availability of budworms increases, whereas the foraging efficiency of their predators decreases. As a result, budworm increase comes closer and closer to exceeding the death toll raised by their predators. Eventually, usually if a period of warm dry weather gives a little extra boost to budworm growth, the balance is tipped, and a runaway cycle of more moths, more larvae, and so on leads to an outbreak. Once this happens in one spot, there is often a domino effect (section 5.2). A local outbreak can spread over thousands of square kilometers and eventually collapse after 7 to 16 years.

After World War II, a campaign to control spruce budworm became one of the first massive attempts to protect a natural resource using pesticides. However, as it turned out later, spraying insecticide only postpones the problem. It allows increasingly large areas to become ripe for an outbreak, leading eventually to much more extensive and devastating events.[35] After a defoliation event, aspen and birch often dominate the regenerating forest, but over a period of 20 to 40 years, selective browsing by moose shifts the forest back to a state dominated by conifers, allowing a new cycle to start.[36]

11.4 The Rise and Fall of Raised Bogs

Raised bogs form in wet climates when shallow open waters are filled by organic matter. At first, this results in semiterrestrial situations, but eventually water-saturated peat accumulates, up to several meters above the water surface (figure 11.6). The formation of such raised bogs appears to be a self-reinforcing runaway process, as described by Nico Van Breemen in a review with the superb title "How sphagnum bogs down other plants."[37] I first summarize how bogs can become a

FIGURE 11.6. Development of a raised bog from a shallow lake. (From reference 37.)

self-propagating system and then show how the life of such bogs may come to an end if atmospheric nitrogen input passes a tipping point.

How *Sphagnum* Bogs Down Other Plants

Raised bogs are almost completely dominated by peat mosses of the genus *Sphagnum*, and one of the central questions in the research on those ecosystems has been how those mosses are able to reach such dominance. The traditional view was that *Sphagnum* is simply well adapted to wet, acidic, and nutrient-poor conditions. However, this appears to be only one side of the coin. The other side is that *Sphagnum* can actually create such an environment if external conditions are within a suitable range. Thus, this little plant is an ecosystem engineer that can create conditions under which its competitors are unable to thrive by making the environment wet, cold, acidic, and nutrient poor. Following Van Breemen,[37] let us look at how each of those aspects is promoted by *Sphagnum* and makes the life of its competitors difficult.

The wet conditions must already be there for the bog to develop. However, once *Sphagnum* peat establishes, it further promotes water stagnation. This happens as burrowing fauna are eradicated and humic substances increasingly clog soil pores. Rainwater is kept in the sponge-like raised layers of the bog, but water can also be drawn up from deeper layers through the capillary spaces formed by overlapping pendent branches. The resulting waterlogged situation results in anoxia that makes such peat a bad rooting environment for most plants.

Sphagnum peat also conducts heat poorly, causing a shorter growing season for vascular plants that root in this cold soil. In contrast, the thin top layer tends to be relatively warm, promoting growth of the superficial active *Sphagnum* layer. The particular physiology of *Sphagnum* makes the bogs notoriously acidic. Although such acidity is not favorable for most other plants, it is not by itself an explanation for the poor growth of vascular competitors, as other, equally acidic soils do have vegetation other than *Sphagnum*. Probably the most important aspect is that *Sphagnum* is an effective competitor for nitrogen and at the same time keeps the ambient concentrations of inorganic nitrogen low. This is a classic recipe for *unstable coexistence* with a competitor (see the appendix, section A.4). Most plants cannot grow at the low nitrogen concentrations caused by *Sphagnum*, but the moss itself thrives. Its effect on ambient inorganic nitrogen concentrations is due to a combination of efficient uptake and slow mineralization.

HOW THE DOMINANCE OF *SPHAGNUM* MAY COLLAPSE

Those modifications of the environment together may explain why raised bogs appear to represent a stable community. The wet, cold, acidic, and nutrient-poor environment created by *Sphagnum* cannot be invaded by trees, shrubs, and other plants. Once the *Sphagnum* community is in place, there seems no way out. At least, this is what is suggested by reconstructions of succession pathways from soil-cores. However, it is becoming clear that the system does have an Achilles' heel. Actually, there are two Achilles' heels. First, drainage can lead to aeration and decomposition, releasing nutrients and irreversibly increasing the permeability of the peat, which then becomes unsuitable as a substrate for *Sphagnum.*

The second weak point is that *Sphagnum* can keep the ambient nitrogen concentrations low only as long as the external nitrogen input is not too high. Leon Lamers and colleagues show how the nitrogen filter effect breaks down once a critical input level is exceeded.[38] Pristine bogs have a low input of nitrogen, mainly through atmospheric deposition and some N-fixation by cyanobacteria. However, atmospheric deposition has increased markedly in many regions. As long as nitrogen is limiting *Sphagnum* growth, an increase in nitrogen input will simply lead to a proportional increase in *Sphagnum* growth, maintaining a constant low ambient level. However, if the nitrogen input rate increases beyond a critical limit (around 15 kg ha^{-1} y^{-1}), the growth becomes limited by other factors. As a result, not all nitrogen is taken up by *Sphagnum,* and the ambient concentrations increase enough to allow other plants to colonize the bog. Once this happens, it may trigger another runaway process. Trees, shrubs, and grasses enhance the atmospheric nitrogen input through canopy interception and depress *Sphagnum* growth by shading. Decomposition is promoted by the increased N-content of *Sphagnum* litter and by the increasing amount of more easily degradable vascular plant litter. Last, the exudation of organic substances and oxygen by roots of vascular plants further stimulates decomposition and mineralization. All this results in further increase of nitrogen concentrations. A simple graphical model may illustrate how such a feedback could cause a critical transition from a *Sphagnum* bog to a vascular plant–dominated system (figure 11.7). The assumptions are that (1) given a certain nitrogen input, soil inorganic N-concentrations will be lower in a system with *Sphagnum* dominance than in a system with vascular plants, and (2) vascular plants need a critical N-concentration for growth that is higher than the concentration needed by *Sphagnum.* The model and its interpretation are entirely analogous to that for the turbidity of shallow lakes (figure 2.9).

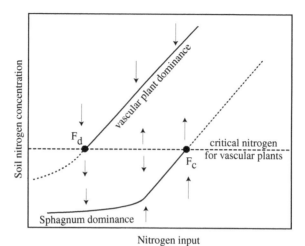

FIGURE 11.7. Schematic model of the competition between *Sphagnum* and vascular plants in a raised bog. At low atmospheric nitrogen inputs, *Sphagnum* utilizes most of this limiting nutrient, keeping soil concentrations low. At higher input levels, other factors become limiting. As *Sphagnum* growth does not increase proportionally with N-inputs anymore, soil concentrations increase. When the critical soil concentration for vascular plants is reached (at F_c), *Sphagnum* dominance is lost and soil N-concentrations jump to a higher level through a range of mechanisms (see text). To restore the *Sphagnum* dominance, one would have to reduce atmospheric input to a level (at F_d) lower than the critical level at which vascular plants took over.

11.5 Species Extinction in Fragmented Landscapes

When it comes to the conservation of endangered species, one of the worries is that their population level may go into free fall if a critical density is reached. This is known as an *Allee effect* (figure 2.3 and appendix, section A.2) and can happen through several mechanisms. A particularly puzzling mechanism is habitat fragmentation. Many populations of animals and plants live in habitat fragments that are dispersed in an otherwise uninhabitable landscape. This is true for the flora and fauna of oceanic islands, but also for the numerous birds, mammals, insects, and plants that depend on fragments of once continuous ecosystems such as natural forest patches in an agricultural landscape. As a result, the populations exist locally in the patches and

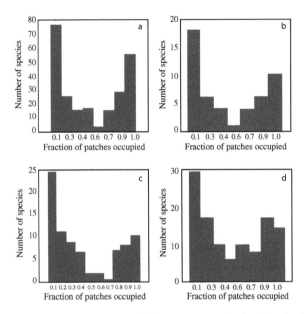

FIGURE 11.8. In areas with fragmented habitats, a species is often found to be either absent from most of the suitable habitat patches or present in most patches. This is apparent from bimodal frequency distributions shown here for (a) human-distributed plants inhabiting small villages surrounded by forest, (b) British butterflies, (c) grebes, and (d) gall wasps on oaks. (From reference 40.)

are coupled by dispersion to form called *meta-population*. Landscape ecologists have studied such situations, focusing mainly on issues such as the risk that species go extinct as a result of excessive habitat fragmentation.[39]

An interesting pattern has been observed in studying the abundance distribution of species in such meta-populations. It appears that the majority of the species occur, at any one time, either in most or all sites suitable for the species, or only in a few sites (figure 11.8). This has been shown for species varying from plants and insects to birds.[40] Such a bimodal pattern is suggestive of the existence of two alternative stable states (section 14.1). Indeed, models of meta-populations predict that we should expect bistability, resulting from a positive feedback between the meta-population size and the local population sizes. A detailed study of butterflies that live on isolated pastures grouped

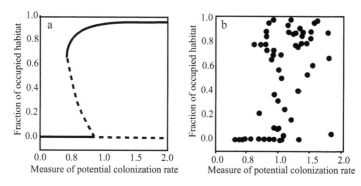

FIGURE 11.9. Models predict that depending on the number of neighboring habitat patches among which colonization can occur, a species will be either present in most patches or absent in all (panel a). Data from Finnish butterflies (panel b) fit the picture remarkably well. (From reference 41.)

on different Finnish islands provides convincing support for this prediction. Each island harbors a meta-population of butterflies that is largely independent of that on other islands, making an extensive comparison possible. The data show that, as predicted by the models, meta-populations tend to go extinct if the number of neighboring patches is too small (figure 11.9).

An obvious implication of this insight is that gradual habitat fragmentation can push species over a threshold at which an unexpected critical transition occurs from a situation in which most patches are occupied to a situation in which the entire meta-population goes extinct. Clearly, it is often difficult to interpret the patterns for isolated species. Species interact with other species that may be affected by the same fragmentation problem. Thus, the total result may not be easy to see from the isolated stories of different species. Rather, one should often focus at *meta-communities* or *meta-ecosystems* to make sense of it.[42] For instance, work on isolated lakes and ponds shows that the loss of fish as a key predator may have cascading effects that allow the entire system to switch from a turbid to a clear state dominated by submerged plants.[43] Therefore, increased isolation of habitat fragments (lakes) may actually promote rather than depress the abundance of many species that prefer the clear water state.

11.6 Epidemics as Critical Transitions

To conclude, it should be mentioned that outbreaks of pests and diseases are perhaps among the best-known examples of critical transitions. Not surprisingly, most of the work has focused on disease outbreaks in human societies. However, the same rules obviously apply to epidemics in other species. Already in 1927, Kermack and McKendrick predicted that epidemics should happen only beyond a critical threshold that depends on the infectivity, recovery, and death rates peculiar to the particular case in question.[44] It was argued that no epidemic can occur if the population density is below this threshold value, and if the population density slightly exceeds its threshold value, an epidemic will occur and reduce the density below the threshold. Although models have become more precise over the years, the original theory has proven largely correct and has important fundamental ramifications. For instance, an epidemic, in general, comes to an end when a threshold fraction of individuals that have recovered and become immune is exceeded. Another important implication is that vaccination can prevent an epidemic even if only some of the individuals are vaccinated.

Virus characteristics of diseases such as flu change continuously over time. Occasionally, a quite novel virus strain arises for which the global population has barely any immunity derived from earlier flu infections. This can lead to well-known pandemics such as the Spanish flu, with its immense death toll less than a century ago. Other diseases change little, and once immune, an individual is protected for the rest of its life. Such typical "child diseases" like measles have rather predictable dynamics that can be described well by simple models.[45] Outbreaks are typically triggered when contact between children intensifies as they crowd in schools. However, they also depend on the fraction of fresh susceptible children who have not yet acquired immunity either through having had the disease or by vaccination. Whenever the threshold is exceeded, an outbreak occurs. In London, for instance, such outbreaks would traditionally happen every two years. However, as in other places, dynamics have become different (and more chaotic) when vaccination became common (figure 11.10). The issue of critical transition from a low incidence to an epidemic is

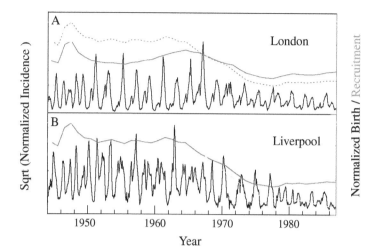

FIGURE 11.10. Dynamics of measles in two English cities. The fluctuating time-series show the normalized measles incidence. The smooth dark curve shows normalized annual birth rate (Liverpool data shown dotted in the London panel for comparison), the lighter gray curve shows annual susceptible recruitment (which has become lower than the birth rate due to vaccination from the late 1960s onward). Together with the mean transmission rate, which depends on seasonal crowding in schools, the recruitment rate of fresh susceptible children determines the nature of the dynamics, which can be mimicked well by simple models. (From reference 45.)

central to work on disease outbreaks in animals ranging from chickens, pigs, and cows to human populations. I will refrain from deeper discussions here, but the literature on the issue is obviously abundant.

One point to note is that although epidemics eventually vanish, they may sometimes tip the system into a permanent alternative state. A striking example is the transformation of boreal forest into lichen plains mentioned earlier. An outbreak of spruce budworm, if it coincides with fire, can cause a transformation that leaves the system in an alternative state that is stable enough to persist for more than a millennium. The collapse of Caribbean coral reef ecosystems is another example. The devastating disease outbreak among sea urchins triggered a transition of the reefs to an algae-dominated state that has persisted until today, decades after the epidemic (section 10.2).

11.7 Synthesis

Clearly, terrestrial ecosystems are easier to study than oceans or past climate dynamics, and there is a large collection of compelling examples of critical transitions between alternative stable states. Nonetheless, many terrestrial ecologists seem reluctant to accept the idea. Why would that be? An examination of most natural landscapes leaves an impression of heterogeneity and diversity. In such a loosely organized colorful patchwork of microsites going through asynchronous phases of succession, one would not expect the sharp shifts that have been described for the seemingly simpler systems of lakes and oceans. Indeed, critical transitions may be uncommon in most terrestrial ecosystems. But so they are in other systems. They are the exception rather than the rule, but not less interesting for that reason. The reviewed cases speak for themselves. Disease outbreaks are among the most obvious examples. They happen when a critical threshold of susceptible individuals is passed. Interestingly, they may incidentally tip the system as a whole across another critical threshold at which point the system slips into another basin of attraction, as illustrated by the boreal forest case. The example of raised bogs illustrates how competitive dominance may be self-reinforcing, driving the entire system in one direction once a threshold is passed. Similar critical transitions in dominance may well happen in a range of other species-poor systems such as heathlands that are invaded by grasses beyond a critical nitrogen loading[46] or boreal forests that can shift to lichen plains. Many of the examples from semiarid systems illustrate how sharp transitions in space or time can arise from positive feedback between plants and their microsite growing conditions. While this results in interesting patterns, the scale is often small. This contrasts strongly with the potentially immense effects of a positive feedback between vegetation and regional climate. The birth of the Sahel–Sahara desert was most likely a critical transition related to such a feedback, and the Amazon basin may have a tipping point leading to a drier alternative stable state in which much of the forest is replaced by savanna.

CHAPTER 12
Humans

The dynamics of human history are as jerky as the dynamics of ecosystems and the climate. Quiet periods occur but are inevitably interrupted by jumps varying from paradigm shifts to revolutions, wars, and state collapses. Although external disturbances such as abrupt climate change have been implicated in more than one example of societal collapse,[1,2] many other dramatic societal changes seem to come without large external impacts. In this chapter, I will suggest that, just as in natural systems, gradual changes occasionally undermine resilience, bringing a social system to a tipping point resulting in a self-propagating critical transition. Such transitions may be unwanted at times. However, as you will see, individuals as well as groups, companies, and societies also have a tendency to get "locked" into a particular attitude or behavior, making them less responsive to changes. "Good" critical transitions may be needed to escape from such situations and other unwanted stable states, such as poverty traps and social disorder.

Before highlighting similarities in dynamics, it is good to note some fundamental differences in the mechanisms that govern dynamics social systems as opposed to "natural" systems. Learning, spread of information, and innovation play an overwhelmingly important role in societies. An important implication is that social systems never really shift back to a state that they have been in before, as ecosystems or the climate system sometimes do. Progressive change does occur in ecosystems through evolution and in the Earth system through continental drift. However, these processes are relatively slow compared to the dynamics of ecosystems or the climate. In contrast, learning, spread of information, and innovation in societies are fast enough to consider each state of a social system fundamentally new. This implies that a partially different set of theoretical frameworks, including ideas such as adaptive cycles that are not captured in simple equations so easily, have been found useful for describing social dynamics (section 4.3). We continuously learn from the past, and hence the future is never really the same.

Even though learning and spread of information sets social dynamics somewhat apart from dynamics in the natural systems, there are obvious similarities too. Classic patterns such as cycles and regime shifts appear in social systems much as they do in natural systems. Certainly, social systems are difficult to model in a quantitative way, and experimental studies on relevant scales are impossible. However, parallels to other dynamical systems are obvious when it comes to nonlinear dynamics. Examples from economics include market cycles (for example, the classic hog cycle because of delays between market-price-driven farmer decisions and final production output) and runaway feedbacks in poverty (the poverty trap) or in the marketing of technological standards (for example, runaway dominance by the QWERTY keyboard or VHS video). Also, the Italian theoretician Sergio Rinaldi has proposed mathematical models to explain how delayed feedback mechanisms can generate stability versus breakdown or cycles in love relationships[3] and a cyclic pattern in corruption as seen in some countries over prolonged periods.[4] Similarly, the dynamics of conflict have been framed in a dynamical systems view.[5] The central idea here is that depending on the circumstances, positive feedback in violence can draw the parties into an irreversible spiral of aggression. The distribution of conflict sizes typically follows a power law,[6] suggesting that self-organized criticality may play a role here, creating smaller and larger avalanches of conflict much as chain reactions of varying size happen in forest fire models and on a growing pile of sand (see section 4.1).

Although the dynamics of social systems are a wide and fascinating area, I will confine myself in the following sections to the issue of inertia and shifts in perception and attitudes, scaling up from the individual mind to the dynamics of groups and societies. This is because I think that this pattern has major implications for the way societies can handle problems such as large-scale shifts in the climate or ecosystems. My thinking about these issues has been shaped by discussions with the sociologist Frances Westley, and this chapter follows the lines of a review that I published with her.[7] I start at the cellular level and proceed to the level of societal dynamics to illustrate again how critical transitions occur across scales. Not surprisingly, uncertainty increases if we go from the cell to society, where the application of theoretical models necessarily becomes more speculative.

12.1 Shifting Cells

On a cellular level, alternative attractors are a common phenomenon. This usually serves a clear purpose. For instance, it is important for cells to "decide" between distinct options over their life cycle. They should become a liver cell or a blood cell, not something in between. Also, during early embryonic development, many distinct "choices" should be made. One of the cases in which a positive feedback leading to bistability has been demonstrated is the dorsal–ventral patterning (which side becomes the back) in fruit flies.[8] Starting with a shallow gradient of a certain type of protein indicating the position of the back, an intracellular positive feedback circuit amplifies this by promoting future receptor binding as a function of previous signaling strength. The result is the development of an ordered sharp stripe of dorsal cells.

   Another example of a cell choice is that of cell suicide through deliberate *programmed cell death*. Apoptosis is one of the main types of programmed cell death. This suicide is carried out in an ordered process and serves a clear purpose. For example, the differentiation of human fingers in a developing embryo requires the cells between the fingers to initiate apoptosis so that the fingers can separate. Apoptosis can also occur when a cell is damaged beyond repair or infected with a virus. Such suicide is a distinct choice. The cell should live or die, not something in between. It appears that this all-or-nothing character of apoptosis is realized through a positive feedback in a system of signaling proteins [9] leading to a bistability of this biochemical reaction. More generally, such bistability in cell signaling pathways serves to filter out noise (irrelevant random fluctuations in the environment) and yet respond decisively if stimuli exceed a certain threshold.

12.2 Shifting Minds

Human minds are notorious too for locking into one of several contrasting states. Mood swings are one aspect. Depression is a complex phenomenon involving not only the mind, but also overall chemical balances and feedbacks involving behavioral patterns and social interactions. Whatever the precise mechanisms are, it is well known that unipolar depression can be a quite stable condition. In contrast, individuals with bipolar depression can shift erratically between widely contrasting states of mania and depression.

On a more subtle level, the mind has a tendency to lock into one of several alternative interpretations of reality. A well-known example is the human tendency to fit visual clues to search images (figure 12.1). It appears difficult to see different interpretations simultaneously, and the resulting "snapping" to one of several alternative interpretations seems to happen on different levels varying from interpretation of pictures to more complex theories and worldviews. Scientists are faced with this problem often in their work. We easily adopt "pet hypotheses" (mine is not difficult to guess), and we unconsciously tend to note the observations that are in line with that hypothesis more than the observations that do not fit the view[10] The same mechanism may play a role in ideology of all kinds, from political to religious beliefs.

One can imagine that the tendency to quickly fit complex observations to search images serves a purpose. It allows efficient recognition of important things such as food items or situations of danger. A vivid account of examples of this classification mechanism in humans can be found in the popular book *Blink* by Malcolm Gladwell.[11] His overview shows that this rapid classification mechanism is much faster

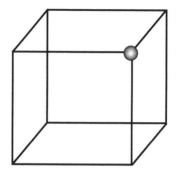

Figure 12.1. Interpretation of images is notoriously ambiguous. You can imagine the marked corner in the Necker cube to be at the front side or at the back side. Once you see one interpretation, shifting to the other one is not easy.

and sometimes also more accurate than the parallel and independent rational track that our brain follows. Similar findings have been documented in studies of how experts make rapid decisions.[12]

12.3 Behavioral Lock-In

Also behavioral patterns have the tendency to lock into a particular mode from which one does not easily escape. Again, this may often serve a purpose. For instance, faced with a predator or enemy, we should fight or flee. Once you start fighting, it makes sense to pursue the task completely, rather than doubting and wobbling between different modes of behavior. Consistency is best.

In the case of aggression, a physiological feedback loop has been detected in rats between aggressive behavior and hormonal production. This feedback can ensure that a burst of aggressive behavior typically does not too quickly fade away. It is probably a quite universal mechanism — you may have noticed this pattern in children's tantrums and in adult's bursts of anger. The aggressive behavior apparently induces a hormonal (adrenocortical) stress response, which in turn promotes a center in the brain to conduct aggressive behavior. This leads to a positive feedback of the controlling mechanisms within the time frame of a single conflict.[13] While this mechanism may contribute to the precipitation and escalation of violent behavior, one can also see that a rapid move to a consistent aggressive (or flight) mode of behavior can give a higher chance of survival than hesitating between different modes of action.

Perhaps the benefits of consistency also explain the curious phenomenon known as the *sunk-cost effect*. Economic theory tells us that prior investment should not influence one's consideration of current options. Only the incremental costs and benefits of the current options should influence one's decision. However, there are numerous examples suggesting that humans deviate from that rational path and can be trapped into a positive feedback between prior investment and behavioral choice. This is called the sunk-cost effect[14] but has been referred to in animal research as the *Concorde effect*, or *Concorde fallacy*.[15] The dim financial prospects of the Concorde were known long before the plane was completed, but the UK and France decided to continue on the grounds that they had already invested a lot of money. In a review of many studies, Arkes and Ayton[14] conclude that the Concorde fallacy does not occur in lower animals, but that many studies demonstrate the sunk-cost effect for humans. For example, a study on American basketball players showed that individuals who cost the team more money were given greater playing time independent of the player's performance.[16] In some cases, at a closer look, seemingly irrational sunk-cost behavior may really be explained by the interest of key individuals to maintain the status quo or prevent loss of prestige. However, there are also indications of intrinsic psychological mechanisms playing a role.[14,17] Among other things, self-justification[18] may play a role, as people often do not like to admit that their past decisions were incorrect. In fact, studies of conversion and brainwashing suggest that beliefs can be an essential part of identity, explaining the large resistance of individuals to such turnover.[19] Whatever the explanation is, adult humans apparently have a tendency to stick to a certain mode of behavior even if this is rationally a bad choice. This lock-in mechanism caused by apparent self-reinforcing adherence to a mode of behavior tends to promote *inertia*, a lack of responsiveness to changes in the environment.

12.4 Inertia and Shifts in Group Attitudes

While individuals have a tendency to lock into one particular inter-pretation or behavior, group dynamics add a second level of inertia. Many studies confirm that public attitude often exhibits sudden rather than gradual shifts, and a wonderful, accessible overview of examples ranging from fashion and smoking, to crime dynamics can be found in the best seller *Tipping Point* by the journalist Malcolm Gladwell.

We cannot conduct controlled experiments to unravel the mecha-nisms on the scale of large groups or entire societies, but experiments in small groups nicely reveal the likely basic mechanism. For instance, early studies in experimental psychology have shown that people's re-sponses to calls for help in emergency situations depend very much on how they read the responses of those around them.[20] If nobody in a group of bystanders acts to help, you are likely to copy that behavior. As a result, groups have a tendency to remain locked in a passive atti-tude, where individuals would already have acted. In hindsight, there is often disbelief in this dynamic. "How is it possible that so many stood by and did nothing?" Ironically, the truth is that it may often be *because of* rather than *despite of* the fact that so many stood by.

The tendency to follow the group's attitude is so strong that people often disavow the evidence of their own senses if other members of a group interpret reality differently.[21] In one famous experiment, indi-viduals working alone to match a line of a given length to one of three on a comparison card did so with less than 1% error. If however, the individual was then placed in a group in which all other members, acting as accomplices in the experiment, chose the wrong line, the same individuals would chose the wrong line in more than one third of the cases.[22]

The tendency to lock into the same attitude implies that groups may often be rigid when it comes to responding to changing condi-tions. Even if there is a general feeling that something would need to be done, it can be surprisingly difficult to get a group out of the grid-lock. In such situations, there is an important role of the "exceptional few" to catalyze tipping points. Some individuals appear to be able to mobilize groups to change because of a combination of factors. For instance, because they are particularly well connected,[23] have high social capital, are by nature innovators or early adopters,[2,4] and /or have the charisma to cause emotional contagion.[25] An absence of such leaders will make a social group as a whole rigid and weak when adaptation to change is required. The same pattern is also observed in paradigm shifts in science, where exceptional minds will reframe a pattern of discovery into new perspectives on old facts.[26,27]

The way in which contingency of attitude and leadership may cause critical transitions can be demonstrated by a simple mathematical model (see the appendix, section A.14, for equations). This model assumes that for each individual, there are two modes of "opinion" (or "attitude") with respect to the question of whether action should be taken against a problem (such as climate change or street crime): passive or active. Individuals adopt an attitude depending on their image of how serious the problem is and how effective it would be to push for regulation. However, their attitude is also affected by peer group *social pressure*. In addition, there is a stochastic component to reflect differences between individuals. In the model, the individuals take an attitude through a cost-benefit argument, assuming a cost of deviating from the overall group tendency (going against peer pressure) and a perceived net utility of taking the positive attitude.

This model can be used to predict in which way the mean public attitude changes in response to a new and slowly increasing environmental problem (figure 12.2). The sigmoidal equilibrium curve is the type of catastrophe fold that we have seen many times and implies in this case that the response to a slow increase in perceived problem size is discontinuous. Most individuals favor a passive attitude until a critical point (F_1) is reached, at which a sudden and fast transition to an active attitude toward combating the problem occurs. The underlying mechanism is that groups tend to stick together when it comes to attitude or opinion. In the model, this is the result of peer pressure, but as argued earlier, the forces keeping a group locked into the same mode are more subtle than that. This "sticking together" implies a tendency to inertia in the face of changing conditions until an avalanche of individual attitude shifts occurs when a threshold is reached. The runaway character of the avalanche is due to the fact that as more individuals shift to the other attitude, the contingency implies that the individuals that are still sticking to the previous attitude become

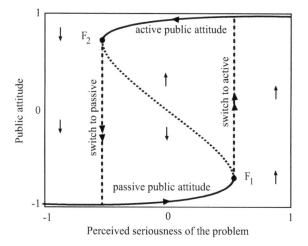

FIGURE 12.2. In societies with little difference among individuals and high peer pressure, the response of public attitude to an increase in perceived problem size is predicted to be discontinuous. When the problem is perceived to be small (and the perceived payoff of taking action is low), the attitude of most individuals is passive with respect to the problem. Society abruptly shifts to a predominantly active attitude (creating political pressure to regulate the problem) when the perceived severity of the problem has grown sufficiently to reach a critical point (F_1). If, subsequently, the severity of the problem is reduced, the active attitude toward regulation remains until another critical threshold point (F_2) is reached where an equally abrupt transition to a passive attitude occurs. The graph is produced from the simple model given in the appendix (see section A.14) by plotting h on the vertical axis and $\bar{A}$ on the vertical axis. (From reference 38.)

increasingly likely to be pulled over too. This dynamic is not unlike the "paradigm shifts" described by Thomas Kuhn,[26] where the accumulation of scientific anomalies in data collected using one perspective results in sudden and radical shift in scientific perspectives and the birth of a new theory that "explains" the anomalies.

Of course, the catastrophe fold also implies that if, subsequently, the size of the problem is reduced, the public attitude still remains in the active mode until another critical threshold point (F_2) is reached, where an equally abrupt transition occurs to a state in which action against the problem is generally considered unnecessary.

The predicted hysteresis depends on the strength of the contingency (peer pressure) and on the degree of variation among individ-

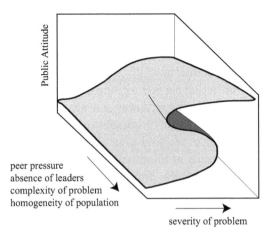

FIGURE 12.3. Observations as well as models suggest that the degree of hysteresis in public attitude toward the need to regulate a problem will be larger in situations with high peer pressure, lack of strong opinion leaders, complex problems, and relatively homogeneous populations. (From reference 38.)

uals. If the peer group effect is weak and individuals differ in their perception, each individual takes an attitude depending on its perceived seriousness of the problem. The resulting average attitude in society smoothly changes with the size of the problem. However, with increasing peer pressure, the mean attitude starts to shift more steeply around a critical perceived size of the problem. Eventually, if contingency is strong enough, the equilibrium curve takes the sigmoidal shape that gives rise to hysteresis and sudden transitions. Decreasing individual variation has practically the same effect in this model as increasing the peer effect. Thus, the strength of contingency and the variance between individuals are *parameters* that can change the response of society to problems from smooth to catastrophic and vice versa (figure 12.3).

The effect of group behavior and the importance of opinion leaders is also illustrated by more elaborate mathematical models that have been explicitly designed to explore the potential effect of strong leaders in social networks. An interesting class of models takes the view that individuals are somehow like magnetic particles that tend to create a magnetic field together but at the same time adjust their positions to be aligned to the field.[28] The emerging image in all lines of model and empirical work in this field is that opinion leaders can precipitate a shift in opinion that would otherwise remain inert to change of external conditions.

The bottom line is that peers in a group have a tendency to become locked into the same set of attitudes and behavioral patterns. This mutual contagion mechanism causes inertia, making it more difficult for the group attitude to respond to new situations (figure 12.3), especially if the complexity of the situation is such that it is not easy to see what is really the problem and how it might be solved. Opinion leaders are highly important in getting out of such a nonadaptive gridlock. This implies chances for innovation and manipulation alike. History provides a wealth of examples ranging from Gandhi to Hitler illustrating that in situations of crisis, the niche for sense making can be filled in very different ways, and the outcome depends greatly on which of the "exceptional few" will catalyze new attitudes.

12.5 Societies in Crisis

Could the tendency to lock into a rigid group mode of behavior also limit innovation at a societal level if a crisis arises? Certainly, the future will be challenging in that sense for humankind in view of the current level of resource depletion and the expected trends in climate, wealth, and population densities. One way to get an idea of how societies may respond to major crises in the future is to look into historical cases. Obviously, humankind has come a long way and countless problems have been overcome. Although this should promote trust in our ability to solve future problems, several cases of dramatic failure stand out that are worth scrutinizing, as they may give some insight into fundamental caveats on the path to adequate societal response to crisis.

Perhaps the best-known class of failures is the collapse of many advanced ancient civilizations facing resource crisis.[2] Clearly, history will not just repeat, as we live in a very different world now in many respects. For instance, local resource scarcity can be resolved by transportation, and technological solutions can greatly enhance resource use efficiency and productivity. Nonetheless, the historical cases suggest some fundamental characteristics of societal response to problems that are deeply rooted in human nature, as revealed by studies of modern human behavior. The main pattern I want to stress here is the tendency to become increasingly rigid and to adhere to old structures and habits as a sense of crisis tightens. Evidence suggests that this may reduce the chance for innovative solutions and much needed change in behavioral patterns.

Perhaps the most striking aspect in the puzzle of the apparent collapse of many ancient civilizations is the power, wealth, and sophistication suggested by the impressive structures they left behind. We know that people were still there after such collapses, but what remained was an apparently simpler society that left little archeological records at all.[29] How is it possible that the same societies that built such structures were unable to avoid falling into the trap that led to complete collapse? Remarkably, there is some evidence suggesting that the inertia in the face of trouble has been because of (rather than despite) the elaborate cities and temples that were built. For instance, archeological reconstructions have revealed that the Anasazi in the U.S. Southwest continued constructing their big cities even during severe periods of droughts, whereas in small settlements in the same region, they abandoned construction during such adverse episodes.[30] This fits with the idea that the sunk-cost effect discussed earlier may have contributed to the apparent inability to respond to crisis in a flexible way. Such a tendency may have made it more difficult for ancient societies to emigrate away from structures in which much had been invested in time to prevent catastrophic collapse in resource crises.[30]

Although the sunk-cost effect is usually linked to overvaluation of established material goods, there is a much broader tendency to cling to worldviews and ways of living as crisis builds up. Jared Diamond has suggested that the fate of the Vikings who colonized Greenland (figure 12.4) might be a stunning example of such an apparent lack of flexibility.[2] The idea is that these Norse people took up the practice of growing domestic livestock that they were used to in their homeland, even though the Greenland climate was very harsh for that. Cows had to be kept inside for most of the year in this climate, and excavated skeletons show that they remained very small probably due to the

FIGURE 12.4. Ruins of a Greenland Viking church. The last written records of the Greenlandic Vikings are from a 1408 marriage in this church. (Photograph by Frederik Carl Peter Rüttel, available free of rights at *http://en.wikipedia.org/wiki/Image:Hvalsey.jpg.*)

poor diet. The Norse used meat of wild animals such as caribou and seals to supplement their diet. Nonetheless, they ran into serious trouble when in the early 1400s the climate dived into an extensive cold period known as the *little ice age.* The Greenland Norse vanished, and the remains suggest that famine was a chief problem. While this may seem explainable in the face of the extreme climatic conditions, the puzzling thing is that meanwhile, the Inuit who were their neighbors seem to have survived. This is probably because the Inuit had superior techniques, especially for hunting seals. The Greenland Norse never copied these techniques even though they lived next to the Inuit for centuries. Instead, they clung to their old way of living. As a result, they starved in the presence of abundant unutilized food resources. Although the story of the Greenland Norse is only one example and might be interpreted in alternative ways (a search of the Internet will provide a number of theories), it seems that the tendency to hold stubbornly to habits that led to great success in the past has brought numerous societies into trouble over history.[2] Failure to change patterns of behavior in stressful situations may well be a deeply rooted caveat for humans. Power structure may further promote rigidity in many cases, as vested interests may make powerful groups reluctant to give up the status quo.

THE EFFICIENCY TRAP

This overview so far suggests that on the cellular level and in individual minds, the benefits of locking into one of several alternative stable modes are often clear, whereas the benefits for locking into a rigid pattern in society are less obvious. Indeed, it seems that locks turn frequently into nonadaptive traps, and one wonders why rigidity in groups can be so common. After all, it seems easy to see the benefits of more adaptive dynamics. Who would deny that critical attitudes and innovative ideas should be always embraced? Nevertheless, reality is different.

Consider an experiment[31] in which groups had to complete complex assignments. In half the groups, the experimenters introduced a "plant" — someone trained by the experimenter to take a critical attitude (play "devil's advocate") in reference to group decisions. The groups with these plants consistently outperformed those without the plants, reinforcing the idea that conflict (within limits) plays an important role in problem solving. Nonetheless, when in the second round of the experiment all groups were asked in secret ballot to eliminate one team member in order to improve performance, all groups who had devil's advocates chose to eliminate them, hence eliminating their competitive advantage. Apparently, few groups recognize the value of diversity and conflict in group problem solving. Is this silly, or might there generally be an advantage to coherent groups dancing to the same beat?

Glancing over different fields of research, it seems almost as if there is a fundamental trade-off between two clusters of properties that we could broadly label as *explorative* versus *efficient (or exploitative)*. In animal studies, such contrasting behavioral syndromes have been known for a long time.[32] However, the trade-off between exploration and exploitation has been studied also in management science, where it has large implications for the way we might run companies.[33] Small innovative companies create new products; to bring them to market, however, they must be launched and reliably produced within reasonable time and cost. This requires increasing efficiency and precision. Waste must be kept at a minimum, volume increased, and price reduced if competitive advantage is to be maintained. The exploration and the exploitation phases require radically different modes of thinking and acting, and indeed require two different organizational cultures.[34] This in part explains the challenge of continuous innovation: it is clearly difficult to simultaneously orchestrate the dynamics of exploration and exploitation. So what to do? Successful large and long-lived companies that depend on continuous innovation have addressed this tension by "encapsulating" creative or explorative units. They often physically separate those research and development teams from the production units and train special managers who can champion and shepherd the innovation process while buffering it from the demands of production. This allows the company to build up a bank of new ideas and products to draw upon in future launches, while simultaneously scaling up in producing and marketing successful initiatives.[35] Others spin off the ideas when they are successful enough, thereby avoiding having to divert energy from creativity to production.[36]

All this suggests that a compromise between the explorative and the efficient modes is typically avoided and therefore is probably a bad compromise in general. It would appear that even minimal amounts of exploration will harm efficiency. One could speculate that this is an underlying reason why in most cases it may simply feel best for group survival and well-being if all members are well in sync and devoted to behavior along the same lines. While this may often be functional for the group, it may also severely limit the adaptive capacity. A striking example is the behavior of groups under siege. In his study of the Bay of Pigs crisis, in which President John F. Kennedy and his group of decision makers made disastrous choices in the interests of group solidarity, Irving Janis[37] coined the now famous term *group think* to describe the propensity of groups in need of new thinking to voluntarily abandon their capacity for problem solving in order to maintain group cohesion. Another example is described in the book *When Prophecy Fails.*[21] In this account, Leon Festinger explores the reactions of a cult that had predicted that the world would end on a certain date and had retreated to a sanctuary to await this momentous occurrence. When the day came and went without the anticipated apocalypse, instead of questioning their prophecy, the group became even more withdrawn and rigid in their thinking.

The tendency to lock into an efficient but nonexplorative mode at times of stress implies the risk of a trap. Let us call it the *efficiency trap*. It limits the chances to escape from a crisis through innovative shifts in strategy. One way of depicting these dynamics is to imagine that in an explorative phase, individuals, groups, or businesses look for an optimum in the *fitness landscape* (figure 12.5). Subsequently, they specialize to become more efficient, raising their particular spot in the fitness landscape further. However, this goes at the cost of the explorative capacity to scan the landscape for alternative good places. This becomes a problem if the landscape gradually changes, causing the originally good spot to end up in a valley of bad fitness. The resulting experienced stress results in further local adaptation, improving the local fitness peak slightly, but also increasing myopia and rigidity further.

CHALLENGES FOR MODERN SOCIETIES

Clearly, the evolutionarily important capacity to lock into consistent modes of behavior may turn into a pathological pattern in human societies if it leads to excessive rigidity in a changing world. Of course, our great advantage is that we can analyze such patterns and learn from them. As we have seen, big businesses have learned to foster innovation with great success. Also, numerous great social innovations have occurred. Indeed, we have learned a lot and keep innovating, so it would be foolish to think that we would be as rigid as the Greenland Norse when it comes to adjusting ways of living in the face of a resource crisis.

Nonetheless, it is hard to deny that societies today remain notoriously slow in responding to new problems.[38] If societal problems or climate or resource crises deepen, will we become innovative enough to find a way out and flexible enough to adjust our patterns of living? Maybe so, but you should be convinced by now that there are also

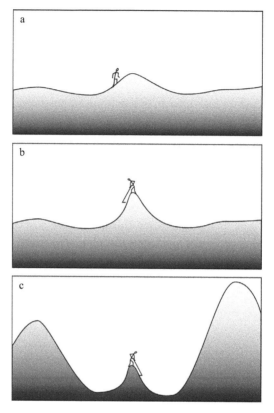

FIGURE 12.5. Illustration of the efficiency trap. If exploration has revealed an optimum in the fitness landscape (panel a), a person, group, or company may shift to an exploitative mode where improved efficiency and specialization enhances the fitness further (panel b). However, in such a behavioral mode, the explorative capacity to scan for alternative options is reduced. This implies the risk of becoming trapped in a situation that may seem optimal from a myopic view but is suboptimal compared to potentially much better places in the changing fitness landscape (panel c).

indications that rigidity in groups sometimes increases in crisis, and smart, integrated solutions may become increasingly unlikely in situations of stress. The dynamics of societal response to problems are obviously driven by a complex set of factors. However, even on this large scale, patterns can be seen that are similar to the ones shown for small groups. For instance, people in nations that live under more stressful conditions (for example, because of poverty or social crisis) exhibit a pattern of increasing dependence on group norms and authority and a decreasing dependency on rationality and individual choice as a basis for making decisions.[39] Clearly, it is rather speculative to extrapolate any such pattern into the future of our societies. However, altogether the evidence suggests that as social and climatic crises in the future become severe on global scales, there is a risk that the resulting stress may tend to result in even more rigid adherence to old patterns. Books have been written on future scenarios, and I will not recapitulate the elaborate visions of numerous aspects and risks here. Nonetheless, understanding the mechanisms of inertia and shifts in nature and society is clearly relevant when it comes to developing new strategies to reduce the risk of future crisis.

12.6 Synthesis

Replicated experiments cannot be performed with societies, and we are far from able to develop accurate mathematical models of social dynamics. Nonetheless, there is abundant evidence for patterns such as inertia and critical transitions that is consistent with the existence of alternative basins of attraction. Also, feedbacks and runaway processes can often be identified that may explain such dynamics. Just as in other systems such as lakes or the Earth system, critical transitions appear to occur in humans on all scales, from the cellular level and individual minds to groups and societies. The tendency to lock into one of several alternative attractors almost always serves an apparent purpose. In cells, it filters out noise and allows a well-defined and consistent behavior once a certain threshold is past. Basically, the same holds for attitudes and behavior of individuals and groups. This functionality should not be surprising, as (unlike nonlinear dynamics of systems such as the climate) it has evolved by selection for fitness. Nonetheless, the tendency to lock into a certain pattern also has a negative side. It comes at the cost of the ability to adjust to new situations. Surprisingly, this rigidity appears to become more pronounced in situations of crisis. Clearly, we would like to prevent particular critical transitions in society such as stock market crashes, collapse of states, and runaway conflict escalation. Some progress is being made on those fronts. For instance, central bank interventions are used to steer the economy away from dangerous thresholds. The other side of the coin is that there may be critical transitions that imply a way out of an undesirable trap. Successful attempts to promote such transitions include the use of microcredit to allow families to escape the poverty trap discussed later. Understanding such dynamics better may help us to find ways to promote our adaptive response to the multitude of problems society faces in a rapidly changing world.

Conclusion:
Critical Transitions in a
Complex World

Looking back on the studies of lakes, the Earth system, oceans, terrestrial ecosystems, evolution of life, and the dynamics of human groups and societies, the emerging image is that although complexity in the details may be overwhelming, pronounced aspects such as cycles and critical transitions between contrasting regimes incidentally stand out of the noise. This happens in all these systems on multiple scales. For instance, in lakes, plankton communities may shift and cycle rapidly, whereas on a slower multiyear timescale, the entire ecosystem may jump from a stable clear condition to a persistent turbid state with very different communities of fish, aquatic plants, and invertebrates. Similarly, in the Earth system, rapid phenomena such as El Niño oscillations and multiyear droughts like the Dust Bowl happen well within human generations, whereas on centennial and millennial scales, climate shifts such as the birth of the Sahara desert and the Younger Dryas cold period stand out. Further, in deep time, dramatic transformations between a tropical and a frozen Earth have occurred. In humans, shifts between contrasting stable states are found in cells, the individual mind, groups, and societies.

It is obvious that our level of understanding of what drives apparent cycles, chaos, or regime shifts differs widely between the systems. Lakes are much better understood than oceans or the Earth system. Clearly,

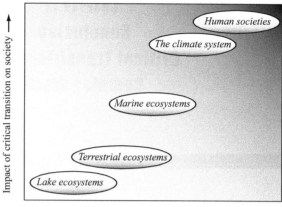

Figure 13.1. Systems in which critical transitions would have the most impact on humans are also the ones in which uncertainty of the models needed for predicting such change is largest.

scale matters. We have learned a lot from experimental manipulations of entire groups of lakes. However, oceans and the climate system cannot be probed by means of experiments on relevant scales. We have to rely on interpretation of effects of natural experiments and on models that put mechanistic insight in different processes together and are tuned until they reproduce observed dynamics. As a result, there is a systematic correlation between uncertainty and scale. The larger and more important the system, the less certain we are about possible thresholds and mechanisms behind transitions (figure 13.1).

Predicting critical transitions is particularly difficult. We are often able to reconstruct the mechanisms behind shifts in hindsight and to identify positive feedbacks in systems that might potentially cause threshold responses or even alternative stability regimes. However, quantitative prediction of such phenomena is limited to smaller systems where we can observe multiple shifts and experimentally probe the thresholds and mechanisms. For instance, we know more or less at which phosphorus loading a shallow lake may shift to a turbid state, as we have seen that happen many times. In contrast, we are not so certain at which freshwater input the thermo-haline circulation might collapse, or if and when part of the Amazon rainforest region may shift irreversibly to a dry savanna state if we keep cutting trees.

Human societies are certainly among the most challenging systems when it comes to understanding and predicting their dynamics. A dazzling feature here is that if we can predict our future, we may also use that information to change our future for the better. This is not a matter of studying humans alone. Societal dynamics are driven in part by change in natural resources and climate, and vice versa. The Earth is simply an integrated complex nonlinear system including us, the climate, and all the ecosystems. Although it may sound ridiculously ambitious and idealistic, the central challenge of our times is in my eyes to understand and predict the main forces that drive the dynamics of the coupled Social-Eco-Earth-System (SEES) and to use this knowledge to carve out the best possible future. Of course, this is a long shot, but at the very least, we can not only attempt to understand what happens on scales from quarks and molecules to individual organisms and jar experiments, but also aim for understanding dynamics on scales that matter so much to society, even though reproducible experimental results are less easily obtained there.

Part III
DEALING WITH CRITICAL TRANSITIONS

Given that nature and societies may go through
sudden and potentially irreversible shifts when
unexpected thresholds are passed, how can we
use this insight to our advantage? This final
part of the book touches on different aspects
of this question. First, I address the problem of
how we might know whether a particular system has the potential for critical transitions
and subsequently the related question of how
we can know whether we are actually approaching a critical threshold. Then, I take a broader
look at how to manage natural resources in the
face of potential critical thresholds. How can
we utilize such systems in a way that would
maximize overall human welfare, and what
stands in the way of taking such an optimal rational approach? Last, I review examples of
preventing unwanted transitions and examples
of the bright side: promoting critical transitions that bring the system to a better state.

CHAPTER 14

How to Know if Alternative Basins of Attraction Exist

You may by now be convinced that insight into the possible existence of alternative stable states and the thresholds for critical transitions is important if we want to manage natural and social systems for the best. However, the case studies illustrate that it is often far from straightforward to interpret observations from real systems in terms of the simple theory presented at the start of this book. Lakes are not so difficult, but how do we obtain reliable information on alternative attractors in societies, oceans, and the climate system? In this chapter, I first review indicators of alternative stable regimes that may be obtained from observational data and then discuss ways to check the alternative stable states hypothesis by means of experiments and models. A condensed version of this chapter has been published earlier as a review that I wrote together with Steve Carpenter.[1]

14.1 Hints from Field Data

Although observational data can always be interpreted in alternative ways, there are particular patterns that do suggest the possibility of alternative stable states. I highlight three types of indications that are commonly associated with the presence of multiple attractors.

Jumps in Time-Series or Regime Shifts

Sudden changes in a system are always an interesting feature, and not surprisingly, various statistical techniques have been developed to check whether a shift in a time-series can be explained by chance.[2] However, it is important to keep in mind that even if a critical transition in a time-series is significant, this does not necessarily imply that it was a jump between alternative attractors. Probably the most common cause of a sudden change is a sudden change in the conditions. For instance, the election of a radical socialist for president in Brazil or a decision of the central bank of the United States to raise the interest rate drastically may cause a marked jump in time-series of all kinds of indicators of the economic system. Similarly, the closure of a dam for a major reservoir may cause a drastic shift in the downstream river ecosystem. Another possible explanation of a sudden shift is that conditions changed gradually but exceeded a limit at which the systems changes drastically but not catastrophically (that is, not a stability shift related to a catastrophic bifurcation). For instance, the onset and termination of a period of ice cover in a lake can be quite sudden, even if temperature develops gradually. Thus, sudden shifts in a time-series of some indicator of the state of a system may often simply be explained by a sudden drastic change in an important control parameter (figure 14.1a) or a control parameter reaching a range where the system responds strongly, even though there is no bifurcation (figure 14.1b). On the other hand, true critical transitions can be caused by a tiny but critical change in conditions (figure 14.1c) and/or by a disturbance pushing the system across the border of a basin of attraction (figure 14.1d). I should stress again that although real stability shifts (two bottom panels) are a distinctly different phenomenon (for example, they can be triggered by infinitely small change and have some irreversibility), there is in fact a continuum of possibilities in the range from linear to catastrophic system responses (more background is given later, in the discussion around figure 14.4).

It may seem impossible to detect from a time-series whether the system went through a real stability shift or instead jumped to one of the mechanisms illustrated in the two top panels of figure 14.1. However, at least in theory, there are some options to sort that out. First, there is a statistical approach to infer whether alternative attractors are

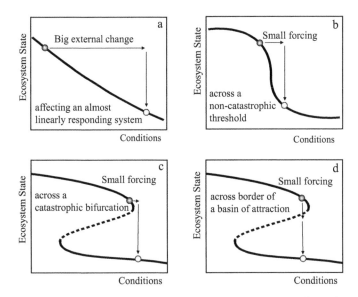

FIGURE 14.1. Four ways in which a sudden jump in a time-series can be explained: (a) a sudden big change in conditions, (b) a small change in conditions in a range in which the system is very sensitive, (c) a small change in conditions passing a catastrophic bifurcation, and (d) a disturbance of the system state across a boundary of a basin of attraction causing a stability shift.

involved in a shift[3] based on the principle that any attractor shift implies a phase in which the system is speeding up as it is diverging from the repelling border of the basin of attraction. Another approach is to compare the fit of contrasting models with and without attractor shifts[4,5] or to compute the probability distribution of a bifurcation parameter.[4] Unfortunately, all such tests require extensive time-series of good quality and containing many shifts.[6] Thus, while jumps in time-series are an indication that something interesting is happening, they are usually not enough to determine whether we are dealing with true stability shifts.

In ecology, much discussion has been devoted to the question of how to map effects of random massive colonization events to stability theory. These happen, for instance, in marine fouling communities[7] that, once established, can be very persistent and hard to replace until the cohort simply dies of old age. It seems inappropriate to relate such shifts to alternative stable regimes,[8] unless the new state can persist through more generations by rejuvenating itself. The latter might be the case, for instance, in dry forests, where adult plant cover is essential for the survival of juveniles except in very rare wet years, which trigger initial massive seedling establishment[9] (see also sections 11.2 and 17.1).

Biological invasions are another phenomenon that is sometimes related to the theory of alternative attractors all too easily. Mostly, the successful invasion of an exotic species is instead a proof that the state of the system in the absence of the species was unstable, in the sense that addition of even a few individuals may lead to a change away from the unstable state at which the density of the species is zero. In other words, many species are absent in many places simply because they have never arrived. Of course, the first stages of an invasion have a strong lottery character, as success depends on the fates of only a few individuals. Hence, chances are higher if the immigration flow is stronger. On the other hand, there can be true alternative attractors involved in invasion dynamics too. In particular, there may be Allee effects (see the appendix, section A.2) implying that the population can establish only starting from a density beyond some critical point.[10] This has important implications for potential ways of preventing or even reversing an invasion. In fact, only in such cases are there good chances to get rid of an invasive species completely. Put simply, one may want to push an unwanted exotic species below its critical density so that it comes into the attraction basin of the alternative stable state in which it is absent. However, in the bulk of the biological invasions, strong Allee effects may not be present, and the state in which the species was absent is in fact unstable. This implies that it is difficult to restore the state without eradicating virtually all individuals.

Sharp Boundaries and Multimodality of Frequency Distributions

The spatial analogue to jumps in time-series is the occurrence of sharp boundaries between contrasting states. For instance, as you have seen, Alpine tree lines can be sharp, and lush kelp forests on rocky coasts can be interrupted by remarkably distinct *barrens* where grazers prevent development of the macro-algae.[11] Similarly, if one samples many distinct systems such as lakes, one may find them to fall into distinct contrasting classes.[12] Statistically, the frequency distributions of key variables should be multimodal (for example, figure 14.2b) if there are alternative attractors. Sophisticated tests are available for multimodality,[13] but again these require rich data sets.[14] Therefore, there is a good chance of concluding that one mode is sufficient even when the data are truly multimodal. On the other hand, significant multimodality does not necessarily imply alternative attractors. There may often be alternative explanations, analogous to those explained in the preceding section on shifts in time-series (figure 14.2a and b): a conditioning factor may itself show a sharp change along a spatial gradient or be multimodally distributed. Also, the system may show a threshold response to a spatially varying factor without having alterative basins of attraction.

THE SHAPE OF A CATASTROPHE FOLD

Part of the difficulty in interpreting jumps in time-series and spatial patterns as indicators for alternative stability domains stems from the problem that we do not know how conditioning factors vary. If one has sufficient data and insight into the role of driving factors, one can push the diagnosis a step further by plotting the system state against the value of an important conditioning factor. Ideally, this produces plots that are directly comparable to the bottom panels in figure 14.1. Statistically, this is not completely straightforward, but one may for instance test whether the response of the system to a control factor is best described by two separate functions rather than one single regression (for example, figure 10.2 and figure 14.2c). Such tests for multiplicity of regression models can be conducted using likelihood ratios, the extra sum of squares principle, or information statistics.[17] If one finds dual relationships, this is suggestive of an underlying hysteresis curve. Still, it may be that a shift in some unknown other control factor has simply taken the system to a different state in which all kind of relationships between variables and environmental factors look different.

In conclusion, one may obtain indications for the existence of alternative attractors from descriptive data, but the evidence can never be conclusive. There is always the possibility that discontinuities in time-series or spatial patterns are due to discontinuities in

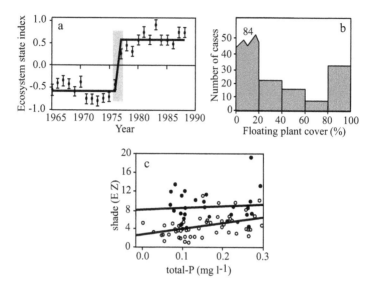

FIGURE 14.2. Three types of hints of the existence of alternative attractors from field data: (a) a shift in a time-series, (b) multimodal distribution of states, and (c) dual relationship to a control factor. The specific examples are as follows: (a) regime shift in the Pacific Ocean ecosystem (modified from Hare and Mantua; see reference 2); (b) bimodal frequency distribution of free-floating plants in a set of 158 Dutch ditches (modified from reference 15); and (c) different relationships between underwater shade and the total phosphorus concentration for shallow lakes dominated by cyanobacteria (heavy dots) and lakes dominated by other algae (open circles) (modified from reference 16).

some environmental factor. Alternatively, the system might simply have a threshold response that is not related to alternative stability domains. The latter possibility is of course still interesting. First, because it helps to know that the system can change sharply if it is pushed across a threshold. Second, because it often implies that under different conditions, true alternative attractors could arise in the same system.

14.2 Experimental Evidence

Experiments can be difficult to perform on relevant scales. However, they are much easier to interpret than field patterns. I discuss here three major ways in which experiments (even "natural experiments") can provide evidence for the existence of alternative attractors.

DIFFERENT INITIAL STATES LEAD TO DIFFERENT FINAL STATES

By definition, systems with more than one basin of attraction can converge to different attracting regimes depending on the initial state. Economists have interpreted the persistence of poverty along these lines (section 2.2), as often poor individuals, groups, or nations stay stubbornly poor, whereas the rich stay rich. In ecosystems, several sets of field observations suggest such so-called *path dependency*. For instance, as explained in chapter 7, otherwise similar excavated gravel pit lakes in the same area of the UK stabilized in either a clear or a turbid state in which they persisted for decades depending on the excavation method.[18] Wet excavation creating initially murky conditions left the lakes turbid. In contrast, if the water was pumped out during excavation and the lake was allowed to refill only afterward, the initial state was one of clear water and such lakes tended to remain clear over the subsequent decades. As always, there might be various alternative explanations for convergence to different endpoints. However, path dependency can be well explored experimentally. The requirement is that one can study a set of replicates of a system that start their development from slightly different states and follow their evolution over time. An example is a study on the competition between floating and submerged aquatic plants (figure 14.3a). The development in a series of buckets incubated with different initial densities of the two plant types was followed. Although the set of initial states represented a gradual range of plant densities, all buckets developed toward dominance by either of the two types, indicating that the mix of the two types was unstable and that dominance by either of the two species represented alternative stable states. Another example of the experimental detection of path dependency comes from a study of plankton communities in small aquariums.[19] Here, it was shown that a different order of colonization from a common species pool may result in alternative endpoint communities that are all stable in the sense that they are resistant against colonization by other species from the pool.

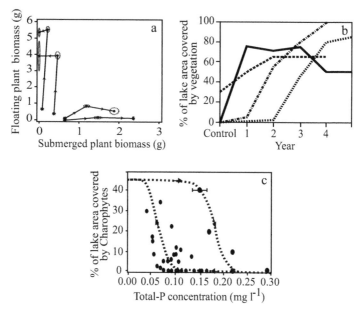

FIGURE 14.3. Three types of experimental evidence for alternative attractors: (a) different initial states leading to different final states, (b) disturbance triggering a shift to another permanent state, and (c) hysteresis in response to forward and backward change in conditions. Specific examples are as follows: (a) path dependency in growth trajectories from competition experiments of a submerged plant (*Elodea*) and a floating plant (*Lemna*), which tend to different final states depending on the initial plant densities (reproduced from reference 15); (b) shifts of different shallow lakes to a vegetation-dominated state triggered by temporary reduction of the fish stock (modified from reference 20); and (c) hysteresis in the response of charophyte vegetation in the shallow Lake Veluwe to an increase and subsequent decrease in the phosphorus concentration (modified from Meijer; see reference 19).

DISTURBANCE CAN TRIGGER A SHIFT TO ANOTHER PERMANENT STATE

Another feature of systems with alternative attractors that can be tested experimentally is the phenomenon that a single stochastic event might push the system to another basin of attraction from which it converges to an alternative persistent regime (figure 14.3b). This is often more practical in a real-life situation than trying to set a large set of *replicate* systems to a slightly different initial condition. Some examples are given in section 17.1—for instance, microcredit, allowing a family to escape from a poverty trap, or a temporary drastic reduction in the fish stock (biomanipulation) to move lakes from a turbid state to a stable clear condition.[21] Lasting effects of single disturbances have also been studied in ecotoxicological research, where the inability of the system to recover to the original state after a brief toxic shock has been referred to as *community conditioning*.[22] All such experiments should be interpreted cautiously. If one wants to demonstrate that the new state is stable, the return of the original species should not be prevented by isolation. Another problem is the potentially long return time to equilibrium, which may suggest an alternative stable regime even if it is just a transitional phase. For instance, the biomanipulated Lake Zwemlust remained clear and vegetated for six years until it started slipping back to the turbid state.[23]

HYSTERESIS IN RESPONSE TO FORWARD AND BACKWARD CHANGES IN CONDITIONS

Demonstration of a full hysteresis in response to slow increase and subsequent decrease in a control factor also comes close to proving the existence of alternative attractors (figure 14.3c). Examples of hysteresis are seen in lakes recovering from acidification[24] or eutrophication[25] and in hemlock-hardwoods forests responding to change in disturbance intensity.[26] However, a hysteretic pattern may not indicate alternative attractors if the response of the system is not fast enough relative to the rate of change in the control factor. Indeed, one will always see some hysteresis-like pattern unless the system response is much faster than the change in the control variable.

In conclusion, experiments are potentially a powerful way to test whether a system may have alternative attractors, but there are important limitations to exploring large spatial scales and long time spans.

14.3 Mechanistic Insight

The natural counterpart to empirical evidence of the overall stability properties of a system is a synthesis of insight into how the system works, in the form of mechanistic models. Such models, formalized as graphs or equations, may explain observed behavior and allow us to extrapolate our findings. Importantly, mechanistic models can also allow us to predict that the system might have alternative stable states, chaotic attractors, or other interesting stability properties, even if those have not yet been detected in its behavior. This may sound like the idea of an overly ambitious theoretician, but it could well be our only resort to address some potentially huge problems. Probing large complex systems such as oceans or societies for alternative domains of attraction is simply not an easy task. Neither field data nor experiments will allow us to really pin down stability properties of such vast, complex systems, yet such insights may be quite important to us. For instance, the thermo-haline circulation of the ocean that ensures a reasonably benign climate to many Europeans and North Americans has never before broken down during interglacial periods. However, we are facing a situation without known past analogue in the Earth climate system, and models suggest that a breakdown of this current in the near future is not impossible. As this would invoke a drastic shift to a much colder climate in some of the economic centers of the world, model predictions are quite relevant. We want such models to become as accurate as possible and should take their outcomes seriously, even though we have no comparable past events to validate this aspect of their predictions.

Mapping mechanistic models to reality is the core of science. It is what allows moving from beliefs to true understanding of the complex world around us. However, the field is also full of pitfalls. This chapter reflects on the power and weaknesses to keep in mind if we attempt to link theory and reality in the analysis of stability properties of complex systems.

Implications of Positive Feedbacks

Positive feedbacks are the source of many of the interesting twists of stability discussed in this book. Threshold behavior, runaway processes, hysteresis, cycles, and chaos all have positive feedbacks as a key component. However, if we spot a positive feedback, we should not jump to conclusions too rapidly. It is tempting, but there is no one-to-one mapping of the detection of a feedback and the resulting stability properties. This is sometimes easy to forget. For instance, positive feedbacks are sometimes interpreted as being almost synonymous to the possibility of switches between alternative attractors.[27] Although this might seem intuitively reasonable, positive feedbacks will not lead to alternative attractors if they are not sufficiently strong.

As an example on a larger scale, take the feedbacks in global warming. There is good evidence that higher global temperatures promote a rise of greenhouse gas levels.[28] For instance, higher temperatures may lead to increased release of CO_2, methane, and nitrous oxide (N_2O) from terrestrial ecosystems and to increased oceanic denitrification and stratification, resulting in nutrient limitation of algal growth reducing the CO_2 sink to the ocean. Although there are also feedbacks that work the other way (for example, higher photosynthetic uptake of CO_2 as the concentration rises), the net result of all these processes is thought to be a positive feedback that will increase the effect of anthropogenic emissions on global temperatures.[29] However, although the Earth system is likely to have alternative attractors on all scales, this particular positive feedback is probably too weak to lead to the possibility of a true alternative stable warm state.[30]

More generally, positive feedbacks tend to create a range of interesting behavior depending on their strength relative to other processes governing the systems dynamics. A modest amplification, such as in the case of the feedback between the Earth temperature and greenhouse gas concentrations, may be common for relatively weak feedbacks. Stronger feedback effects can lead to a relatively sharp response around certain thresholds. Even stronger feedback effects then cause alternative attractors with all the dynamic consequences. As the strength of a feedback often varies with certain system properties, it can be that the same type of system may have alternative stable states under some conditions, while responding in a threshold fashion or smoothly under other conditions (figure 14.4). For instance, as explained earlier, a positive feedback between water clarity and the growth of submerged plants is present in most lakes. However, its effect is stronger if the lake is shallower, as the plants can cover the complete lake bottom there, and their effect on the shallow water column is relatively strong. Hence, this positive feedback is likely to lead to alternative stable states only in shallow lakes where the effect of plants on water clarity can be strong (section 7.1). Similarly, public attitude may change gradually in diverse communities with little social control but

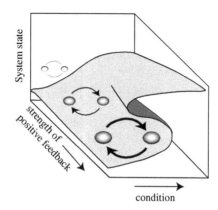

FIGURE 14.4. A positive feedback can lead to modest amplification, threshold responses, or true alternative stable states, depending on its strength.

shift sharply between alternative modes in situations with strong peer pressure (section 12.4). Last, regional climate systems can show hysteresis or not depending on the strength of the feedback effect between vegetation and precipitation (sections 11.1 and 5.2).

In summary, the insight that a positive feedback exists in a system is valuable to alert us to the potential for interesting behavior. However, to be able to predict the consequences of the feedback, we need to take a more quantitative approach and analyze its interaction with other processes in the system.

This brings us back to the issue of the role of mathematical models in science. They are crucial to help in the diagnosis of mechanisms that drive complex systems, but their interpretation has some tricky sides. Discussion of the long history of confusion over the question of what exactly models can tell us is beyond the scope of this book, but it is important to reflect on some of the deep issues that fire debate over and over again.

MINIMAL MODELS AND THE PROBLEM OF HYPOTHESIS TESTING

All of the graphical models in this text and all of the mathematical models in the appendix are so-called *minimal*, or *strategic*, models. Such models focus on a minimal set of mechanisms needed to produce a certain behavior.[31] They are often useful to explore mechanisms that are just too intricate to grasp from common sense alone. However, a drawback of minimal models is that they necessarily leave out many potentially important aspects. They are a step beyond word models such as simple descriptions and diagrams of a positive feedback, but they lack quantitative realism. Adding more and more aspects to strategic models can improve their realism. However, one eventually ends up with big simulation models. As discussed later, such models intended to provide quantitative predictions are much criticized, as because of their complexity, it becomes difficult to determine whether they make sense.[31] Perhaps the most important point to keep in mind with models is that they are basically nothing more than hypotheses. As with any hypothesis, the question of whether they are right or wrong has some problems that have been discussed over and over in the history of science.

First, as Thomas Chamberlin[32] remarked more than a century ago, we should be alert to "the imminent danger of an unconscious selection and of a magnifying of phenomena that fall into harmony with the theory and support it and an unconscious neglect of phenomena that fail of coincidence." While keeping an eye open for alternative explanations is important in all science, it is especially essential in studies of complex systems such as societies or environmental systems. In fact, the classic ideas about hypothesis testing are of little help here. A small excursion to philosophy of science may clarify this.

The classic way of doing proper science is the hypothetico-deductive approach. The main ideas were advocated as early as 1620 by Francis Bacon in his *Novum Organum* and were elaborated later by the influential science philosopher Karl Popper. The idea is to (1) devise some alternative hypotheses; (2) do a crucial experiment, the outcome of which will, as nearly as possible, exclude one or more of the hypotheses; (3) recycle the procedure, making subhypotheses or sequential hypotheses to refine the possibilities that remain; and so on.

While this sounds solid, there are fundamental reasons why such a clean approach is of limited use in ecology[33] and other branches of science that focus on complex systems. A key problem is that competing hypotheses to explain observed phenomena are usually not mutually exclusive. Several independent mechanisms often contribute to cause a phenomenon that could in theory also be explained from each mechanism alone. For instance, fisheries and climatic variation often interact to drive shifts in marine ecosystems. The collapse of Peruvian anchovy (see section 10.1) has likely been pushed by heavy fishing pressure. However, analysis of fish scales in sediments suggest that cycles of boom and bust in anchovies and sardines have appeared long before substantial fishing started, and swings in the system of oceanic currents are likely to be highly influential to these fish populations.[34] The mix of climatic and human impacts has fueled fierce discussions, as some have argued to "let fisherman off the hook" and simply blame climate vagaries in the search for the cause of fisheries collapse.[35]

The core of the problem is that we cannot identify one isolated causal factor in many cases. The cause will usually be a mix that changes from case to case, and from time to time. In practice, this means that neither "clean" controlled experiments,[36] nor minimal models[31] tell us much about the suite of mechanisms that drives real dynamics in complex systems. To stick to the models, even if a model is based on reasonable assumptions and its behavior mimics the patterns in the real system very well, it might be that these patterns are caused by something else in reality. The fact that the modeled mechanism can be shown to operate in reality is not a sufficient basis either for concluding that this mechanism offers the appropriate explanation in that specific case. This is because the modeled mechanism may well be acting in concert with other, possibly more important mechanisms in reality. Thus, simple models (as well as small-scale controlled experiments) can be powerful ways to demonstrate that a certain mechanism can in principle explain a certain phenomenon, but not to demonstrate the relative importance of that mechanism in real situations.

REALISTIC SIMULATION MODELS

The solution to escape from the limitation of qualitative reasoning and minimal models seems straightforward. One simply needs a model that incorporates all important processes in a well-balanced and quantitative way. For several chemical and physical problems on limited scales, such models work relatively well. For instance, the effect of discharge of cooling water of a power plant on the temperature of a river can be computed well. The effect of industrial wastewater loaded with various chemicals on the concentration of these chemicals in sediments, water, and organisms can also be computed reasonably well. In contrast, the effects of CO_2 emissions on the climate and the ocean currents are much more difficult to compute.

Predictive simulation models also work surprisingly poorly for ecological problems. In the early 1970s, there was still a great optimism about the possibilities of constructing detailed simulation models for predicting the behavior of ecosystems. Cooperation of groups of experts on all relevant biological and technical subtopics led to models integrating the available knowledge as much as possible. A lake model,[37] constructed as part of the International Biological Program research, is a good example of this approach. The model contains a diverse spectrum of components including several fish species, algae, zooplankton, aquatic macrophytes, invertebrates, and nutrients, formulated in twenty-eight differential equations. The idea of such modeling approaches was that in the course of the modeling process, lacking information could be identified and filled in after additional experimental research. The latter, however, appeared an impossible mission. The number of parameters in those complex models is very large, and the value of many parameters cannot be determined within a reasonable amount of time, if measurable at all.

The common solution is to estimate the remaining parameter values by fitting the model predictions to field data, so-called *tuning*. A wide array of sophisticated techniques is available for this purpose, and often an impressively good fit is obtained. However, this success is illusive. The problem is that a certain system behavior can often be produced from many different parameter settings. Therefore, tuning of complex ecological models easily leads to good results for the wrong reasons. A good fit does not guarantee any realism of parameter values or model structure. As a consequence, such simulation models have basically the same problems as empirical input–output models. The assumed causal relations need not be true, and therefore extrapolation to new situations can easily lead to nonsense predictions.

Although the bottom line is that true validation of large simulation models is simply impossible,[35] we still need them. One simply cannot conduct appropriate experiments with large systems, such as oceans, societies, and the atmosphere. In these situations, our best hope is a combination of multi-interpretable records of past behavior with *plausible models* of the system. One helpful approach is to use different models in parallel. Each model is a "lie" in the sense that it is an imperfect representation of reality. However, if various independent models coincide in predicting alternative attractors, one can adopt the philosophy that "the truth is the intersection of independent lies."[39] Also, one may get some estimate of uncertainty from such comparisons. For instance, the chances that the thermo-haline circulation in the ocean may break down have been evaluated by comparing an ensemble of predictions from different models.[40] Nonetheless, such sets of models are not really independent. They are often based on largely the same assumptions. Also it is difficult to avoid some bias resulting from the tendency to work toward desired outcomes. This can be a subtle matter. For instance, one may be inclined to check the model for mistakes after a surprising outcome but be much less inclined if the simulated patterns seem fine.

14.4 Synthesis

In conclusion, there is no silver-bullet approach to determining whether a system has alternative basins of attraction. Observations of sudden shifts, sharp boundaries, and bimodal frequency distributions are suggestive but may have other causes. Experiments that demonstrate hallmarks such as path dependency and hysteresis are much more powerful but can be done only on small, fast systems. Models that formalize mechanistic insights are essential to help improve our understanding of complex systems but remain difficult to validate. Clearly, our best bet is to build a case carefully, using all possible complementary approaches, and interpret the results wisely. It is important to remember that although it is difficult to determine whether a threshold exists, decisions must often be made. It then becomes critical to consider the costs and benefits associated with different possible mistakes, including the assumption that there is no threshold, whereas in reality, there is one. We will come back to the issue of management decisions later, but before becoming too focused on the difficulty of proving the existence of critical thresholds, it is important to look at the other side of the coin. If the potential costs of a potential critical transition are very large, it can be well defended that the burden of proof should be on the other side: we may proceed with business as usual only if it can be proven that there are no critical thresholds ahead. Obviously, this is equally as challenging as showing the opposite. Not surprisingly, rational economic analyses suggest that a highly cautious approach will often make most sense in systems where thresholds are suspected.[41]

CHAPTER 15

How to Know if a
Threshold Is Near

The bottom line in the preceding chapters is that although the existence of alternative domains of attraction can well be demonstrated in small controlled systems, we will inevitably remain much less certain about stability properties and critical thresholds in large complex systems such as societies, oceans, and regional climate systems (figure 13.1). Yet in such large systems, occasional sharp shifts have happened in the past, and in view of the rapid change in driving variables such as greenhouse gas concentrations, human population densities, and global nutrient cycles it seems quite unlikely that occasional stability shifts would not happen again in the future. Given that we may never have quantitatively accurate models of large complex systems, it would be very useful to have empirical indicators that can tell us whether we are approaching a critical transition. Clearly, regime shifts that are merely due to some unanticipated big external impact will always remain unpredictable. However, could it be possible to detect somehow that the system is losing resilience and approaching a threshold? At first glance not, as typically the state of the system appears to change little before the threshold is reached. Since a system always changes in one way or another, the kind of change in the state prior to a critical transition will be difficult to distinguish from normal trends and fluctuations. However, on a more subtle level, there may be generic indicators that signal an upcoming shift. Of course, in specific systems, particular changes may be used as early warning signals. For instance, before shallow lakes shift to a turbid state, changes in the structure of the fish community and an increased growth of the algal layer covering submerged plants may be seen.[1] However, the question we address here is whether there could be early warning signals of upcoming shifts that can be used even in systems that we do not understand well.

As you will see in the following sections, there may indeed be such universal early warning signals. The secret to finding these is to make use of the fact that even if the equilibrium of the system may sometimes change little as a bifurcation is approached, the stability landscape does change. Since the stability landscape reflects how the system moves when it is out of equilibrium, this implies that the nature of fluctuations around the equilibrium may carry information about the possible vicinity of thresholds.

15.1 The Theory: Signs of Upcoming Transitions

We first look at early warning signals as predicted from theory and explore by means of simulation models. The mathematics of some of the background is rather intricate, and the work in this field is still under development as I write this. I will not go into the technical details here. Instead, I offer an intuitive explanation of the main contours that are now emerging.

SLOWER RECOVERY FROM PERTURBATIONS

The most important clues that we can obtain to sense whether a system is getting close to a critical threshold are related to a phenomenon known in dynamical systems theory as *critical slowing down*. In technical terms, what happens is that in bifurcation points, the *dominant eigenvalues* characterizing the rates of change around the equilibrium are reduced to zero. This implies that the system becomes very slow in recovering from small perturbations. This is best seen through an example. For instance, consider a population with alternative stable states that is pushed by some factor to come close to the bifurcation

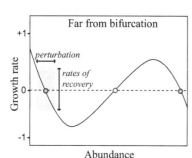
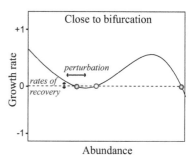

FIGURE 15.1. As a rule, rates of recovery upon small perturbations are smaller if a system is closer to a bifurcation. This example represents the growth rate of a population as a function of its abundance. The solid dots are stable points (corresponding to the solid part of the curve in figure 15.2), whereas the light dot is an unstable equilibrium (corresponding to the dashed part of the curve in figure 15.2). This population has a strong Allee effect, implying that it moves to the left-hand stable point if it starts at an initial abundance below the critical threshold represented by the open dot. The bifurcation happens as the stable point and the unstable point collide (righthand open dot in figure 15.2).

where the unstable point (marking the border of the basin of attraction) moves close to one of the stable points (figure 15.1). Eventually, at the bifurcation, the stable and unstable points touch and disappear. As the system gets closer to that bifurcation, the slope around the stable point becomes less steep, and hence the rates of change around the equilibrium become small. Therefore, as a system moves gradually toward such a critical threshold, recovery from slight perturbations will gradually become slower and slower.[2] The most straightforward implication of this phenomenon is that we may in principle probe the distance of a system from a threshold by studying the recovery rate upon small experimental perturbations (figure 15.2). Since it is the rate of change close to the equilibrium that matters, such perturbations may be very small, and therefore pose no risk of bringing the system over the threshold. Phrased loosely, this is a way to probe the fragility of a system without destroying it.

Obviously, recovery from larger perturbations may also be delayed if they bring the system close to the unstable equilibrium that represents the tipping point. However, thinking of practical applications, such large-scale perturbation experiments for testing the width of the

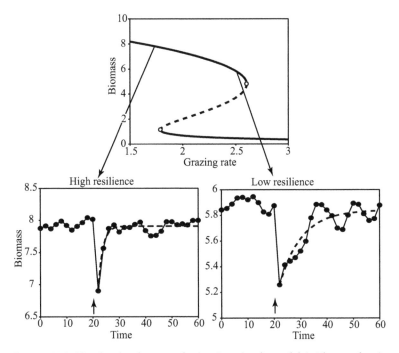

FIGURE 15.2. Simulated pulse perturbation in a simple model (with a stochastic component) showing the difference in recovery rates between a situation far from a bifurcation (bottom-left panel) and a situation close to a bifurcation (bottom-right panel). (From reference 2.)

basin of attraction may be dangerous, as the risk of inducing a large-scale catastrophic shift is large. It is therefore interesting that, unexpectedly, even the response to small perturbations that do not bring the system in the vicinity of a tipping point may somehow reflect the size of the attraction basin. Nonetheless, there are also limitations to the approach. For instance, although the decrease of recovery rate close to a threshold is generic, the precise relationship between recovery rate and the width of the basin of attraction will be specific to any particular system. Thus, absolute values of recovery rates cannot tell us how close to a threshold we are. Indeed, it makes little sense to try to interpret differences in recovery rates between a forest and a plankton system in terms of their vicinity to a threshold. Plankton recovery rates will always be faster, even if the planktonic system is close to a critical threshold. In contrast, differences in recovery rates are informative. For instance, if we want to compare the width of the basin of attraction of two clear shallow lakes or to monitor ecological resilience of a lake as nutrient loading increases, differences in recovery rate upon small perturbations are likely to be meaningful. An important possibility is that in spatially extensive systems, one may perturb the system locally and measure the rate at which the experimental patch returns to equilibrium. For instance, models show that recovery time upon local perturbations should increase in fragmented populations approaching a threshold for global extinction.[3] Recovery from experimental local eradications could thus in principle reveal the resilience of the system as a whole.

INCREASING AUTOCORRELATION

For most systems, it will be impractical or impossible to monitor them by systematically testing recovery rates. There is, however, a smart way around this problem. Most systems are permanently subject to natural perturbations. In such systems, as a bifurcation is approached, one should expect that the slowing down leads to an increase in autocorrelation in the pattern of fluctuations.[4] Loosely phrased, as slowing down causes the rates of change in the system to decrease, its state at any given moment becomes more and more like its past state. Therefore, lag-1 autocorrelation (correlation of the time-series to itself shifted one time-step back) can be interpreted as slowness of recovery in such natural perturbation regimes. Indeed in models that have been tested,[5] a clear increase in autocorrelation can be seen as they approach a catastrophic bifurcation where a critical transition occurs (figure 15.3). Just like critical slowing down, the time-series-based indicators are relative rather than absolute measures of the distance to the threshold.

INCREASING VARIANCE

Variance in fluctuations may also often increase as a critical transition is approached.[6] While the explanation of slowing down is straightforward, variance may be affected in more complex ways. Intuitively, if

Runaway Glaciation

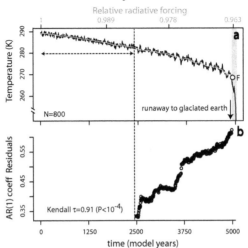

Thermo-haline circulation

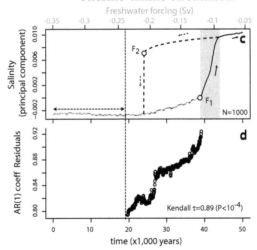

FIGURE 15.3. Simulated time-series of critical transitions generated by two models of climate systems that are slowly driven across a bifurcation. Analysis of the detrended time-series reveals that the shifts are preceded by an increase in slowness estimated as autoregression coefficients (bottom panels). The gray bands identify the transition phases. The arrows mark the width of the moving window used to compute autocorrelation. (From reference 5.)

the system becomes increasingly slow in returning to equilibrium, random perturbations pushing the state around will lead to a pattern that will increasingly look like a random walk, implying increasing variance. However, if one imagines the response of a system to fluctuations in conditions (rather than perturbing the state), it is not directly obvious what should happen as a threshold is approached. As a first approximation, suppose that our dynamical system is simply tracking variations in the position of its attractor driven by environmental fluctuations instantaneously. This is equivalent to saying that the system is infinitely fast in settling to its equilibrium. In that case, the equilibrium curve of the system plotted against an environmental variable (condition) gives us a direct clue to the way in which it will follow environmental fluctuations. As the slope of such curves usually becomes steeper close to a threshold, environmental fluctuations will tend to become amplified by the system as the threshold is approached (figure 15.4). Therefore, if the regime of environmental fluctuations remains the same, one should indeed expect that the variance in the fluctuations of the systems state may increase as a threshold is approached. However, no system will settle immediately to a changed equilibrium, and in fact the increased slowness may increasingly limit the capacity of the system to track the fluctuations in conditions as the threshold is approached. This could in principle reduce variance, as the system becomes too slow to follow the fluctuations in conditions. Nonetheless, increase of variance can be shown to announce a critical transition in a range of models.

FLICKERING

There is another, somewhat different phenomenon that may be seen in time-series in the vicinity of a catastrophic bifurcation point, sometimes referred to as *flickering* or *stuttering*. This happens if stochastic forcing is strong enough to start throwing the system back and forth between the basins of attraction of two alternative attractors as a system enters the bistable region in leading to the bifurcation. Such behavior can be considered an early warning too, as such a system may well shift permanently to the alternative state if the underlying slow change in conditions goes on moving it eventually to a situation with

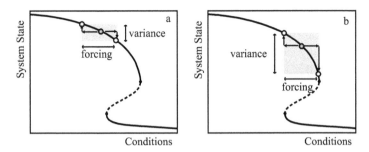

FIGURE 15.4. If one assumes that a system responds infinitely quickly to fluctuations in environmental conditions, it follows that as a fold bifurcation is approached, the variance of fluctuations in the state of the system (for a given forcing "noise" amplitude) becomes larger.

only one stable state. Flickering has been shown for instance in models for lakes going through a transition from clear to turbid and also results in increased variance and autocorrelation,[6] albeit through different mechanisms than the ones discussed earlier, when the dynamics stay close to one of the attractors before the transition happens. A particular case of flickering is known as *stochastic resonance*. This happens if stochastic perturbations are combined with a slow periodic change in conditions. The resulting behavior is a series of more or less periodic shifts before the final transition to the alternative stable state happens. This has been shown, for instance, in a model of the thermo-haline circulation.[7]

INCREASED SPATIAL COHERENCE

In addition to subtle signs in time-series, there are particular spatial patterns that may arise prior to a critical transition. Many systems can be seen as consisting of numerous coupled units that all tend to take a state similar to that of the connected units. For instance, financial markets affect each other, and neurons in networks are affected by connected neurons. Also, the attitude of individuals toward certain issues is affected by the attitude of peers (section 12.4), and persistence of species in habitat patches in a fragmented landscape depends on their presence in neighboring patches from which they can be recolonized (section 11.5). Models suggest that in such systems, phase transitions may occur much as in ferromagnetic materials where individual particles affect each others "spin."[8] As gradual change in an external forcing factor (for example, magnetic field or global economic situation) drives the system closer to a transition, the distribution of states of the units in such systems may change in characteristic ways. A quite generic feature appears to be the tendency toward increased spatial coherence, measured as increased cross correlation or resonance among parts prior to a critical event.[8]

Particular changes in patch patterns can also signal an upcoming transition, but these tend to be specific to certain classes of systems. For instance, patterns become scale-invariant (fractal) as a phase shift is approached in classical models of phase transitions such as the Ising models and related models mentioned earlier.[8] In contrast, in models of systems governed by local disturbance (for example, grazers foraging locally on vegetation patches), the opposite may happen: fractal geometry is found for a large parameter range but gets lost because patches of larger scales vanish as a transition is getting near.[9,10]

Last, in systems with self-organized regular patterns, critical transitions may be announced by particular spatial configurations. For instance, models of desert vegetation show that as a critical transition to a completely stable barren state gets nearer, the vegetation is increasingly characterized by small isolated spots, implying that this may be interpreted as early warning signal (section 4.1).

15.2 Precursors of Transitions in Real Systems

It is one thing to show early warning signals in models, but quite more challenging to pick up such signals in real systems approaching a threshold. Nonetheless, some significant advances have been made. For instance, the truncated size distribution of vegetation patches is seen in drylands that are supposedly close to losing all vegetation.[9] Certainly, one of the most spectacular findings when it comes to early warning signals in real complex systems is the discovery that in several ancient abrupt climate shifts, an increase in autocorrelation happened well before the sudden transition (figure 15.5). Interestingly, this pattern could be shown for abrupt climate shifts ranging from the birth

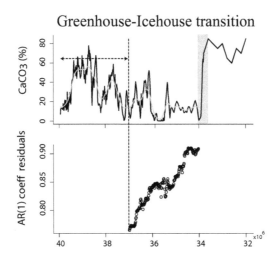

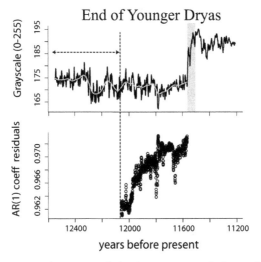

FIGURE 15.5. Increase in autocorrelation in reconstructed climate dynamics preceding the end of the greenhouse Earth and the end of the Younger Dryas. The arrows mark the width of the moving window used to compute autocorrelation. (From reference 5.)

of the Sahara, to the abrupt end of ice ages, and even to the green-house–icehouse transition 34 million years ago. Detecting these patterns required quite a bit of data processing.[5] For instance, slow trends needed to be filtered out so that only the relatively fast fluctuations that supposedly reveal the intrinsic rates of change in the climate system are analyzed. Unexpectedly, variance does not seem to increase prior to the ancient climatic shifts. In contrast, a time-series of a lake shifting from turbid to clear has been shown to contain evidence of an increase in variance prior to the shift, but not of rising autocorrelation.

Empirical work on early warning signals in large systems such as the climate, society, or ecosystems is only beginning to emerge. However, more work has been done in physiology to predict sudden transitions such as epileptic seizures and the onset of heart fibrillation. The underlying dynamical phenomena differ somewhat from those that seem to dominate the climate and ecosystems, where despite all complexities, a classical fold bifurcation may somehow be the underlying structure. The transitions in the heart and brain function appear to be related to changes in the synchronization between oscillating cells and thus correspond to phase locking and resonance phenomena. Nonetheless, similar early warning signals can be picked up in these systems too.

Epileptic seizures happen if neighboring neural cells start firing all in synchrony. Before the seizure becomes noticeable, several characteristic changes in neural activity have been shown to occur. For instance, minutes prior to an epileptic seizure, variance in the electrical signal picked up by an electroencephelograph (EEG) may increase.[11] More subtle changes (reduced *dimensionality* of the signal) occur up to 25 minutes prior to epileptic seizures, reflecting a continuous increase in the degree of synchronicity between neural cells.[12] Also, hours before the seizure, mild energy bursts can occur, followed by frequent symptomless seizures too small for the patients to recognize.[13] This resembles patterns of flickering where smaller transient excursions to the vicinity of an alternative state precede the upcoming major shift. Heart failure is another major area of research. In a particular fatal form of heart failure, it has been recognized that there are early warning signals to be seen in the rhythm, hours prior to the actual event.

In this case, a decrease of the *point correlation dimension* of the heart rhythm appears an accurate predictor of fatal heart failure.[14]

15.3 Reliablility of the Signals

Obviously, detecting early warning signals in data is much more difficult than showing that they work in models. There may be false positives as well as false negatives for a range of reasons. False negatives are situations where a sudden transition occurred but no early warning signals could be detected in the behavior before the shift. This can have different explanations. One possibility is that the sudden shift in the system was not preceded by a gradual approach to a threshold. For instance, the system may have remained at the same distance from the bifurcation point but been pushed to another stable state by a rare extreme event. Also, a shift that is simply due to a fast and permanent change of external conditions (figure 14.1a) cannot be detected through early warning signals. A second class of false negatives may arise from the statistical difficulty of picking up the early warning signal. For instance, detecting increased autocorrelation requires quite long and good time-series. A third difficulty arises if the external regime of perturbations changes over time. This may distort or counteract the expected signals. False positives (or false alarms) arise if a supposed early warning signal is not due to approaching a bifurcation. This may again result from a confounding trend in the external regime of perturbations (increasing in variance or autocorrelation) or may happen just by chance.

Importantly, all of the supposed early warning signals also appear to occur as the system approaches a threshold that is not related to catastrophic bifurcations (cf. figure 14.1b). This has been shown for slowing down[2] and will therefore probably be true as well for autocorrelation and variance. Also, spatial signatures of upcoming transitions such as truncation of the size distribution of patches can happen at noncatastrophic thresholds.[9] Nonetheless, as a rule, noncatastrophic thresholds are related to the more spectacular catastrophic ones, and in fact, systems may move from one type to another type of threshold (figure 14.4). Therefore, even though the early warning signals do not clearly distinguish between the two types of thresholds, they will usually signal an important event coming up.

15.4 Synthesis

Perhaps the most striking aspect of the early warning signals found so far is their generic character: flickering appears to happen prior to epileptic seizures, but also in the thermo-haline circulation before it switched on permanently after the last glaciation; increased auto-correlation may signal critical slowing down before all kinds of climatic transitions, but also in ecosystems; and increased variance of fluctuation may be a leading indicator of a major transition in a lake, as well as of an epileptic seizure. The explanation of this generic character is probably that indeed these transitions are all somehow related to bifurcations, where universal laws of dynamical systems rule the pattern. In fact, the reasoning can be reversed. For instance, we may consider the slowing down of climate fluctuations before a range of ancient shifts as evidence for the existence of tipping points.

The root of most of the early warning signals appears to be the critical slowing down in bifurcation points. As a result, systems become in a sense lethargic. This may again be intuitively derived from stability landscapes (figure 15.6). Close to the bifurcation, the slopes become less steep, implying that recovery upon perturbations becomes slower. As a result, in a fluctuating environment, the state today will look more and more like the state yesterday, and variance will increase. Those early warning indicators share two limitations:

- The indicators will also signal a threshold that is related not to a bifurcation but merely to a strong sensitivity of the system around a certain critical value.
- The absolute value of the indicators is difficult to interpret. (Instead, one should study how they change.)

In any case, generic early warning signals will remain only one part of the toolbox we have for predicting critical transitions. In systems such as lakes where we can observe the transition in many cases, we can empirically determine where the thresholds are in terms of pollution

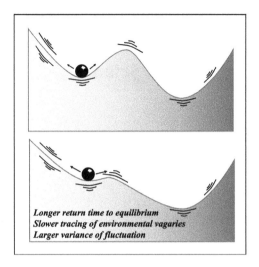

Longer return time to equilibrium
Slower tracing of environmental vagaries
Larger variance of fluctuation

FIGURE 15.6. As the system gets closer to a bifurcation, the size of the attraction basin shrinks, but the basin will usually also become shallower. This implies that the intrinsic speed with which the system moves toward its equilibrium decreases. Therefore, the return time to equilibrium upon a small disturbance (or "displacement of the ball") will tend to become longer. Similarly, the system may become slower to track the equilibrium if the position of the equilibrium varies because of environmental fluctuations. This implies that the fluctuations in the system become more autocorrelated—that is, the signal becomes redder.

or other factors. In other systems, we may want to build the best possible simulation models based on mechanistic insight in an attempt to predict thresholds. Unfortunately, systems such as the climate are unique and give little opportunity of learning by studying many similar transitions. Also, we are far from being able to develop accurate models to predict thresholds in most complex systems ranging from cells to organisms, ecosystems, or the climate. We simply do not understand all relevant mechanisms and feedbacks sufficiently well in most cases. The generic character of early warning signals we discussed is reason for optimism, as they occur largely independent of the precise mechanism involved. Thus, if we have reasons to suspect the possibility of critical transitions, it may well be wise to monitor for early warning signals.

CHAPTER 16

The Winding Road from Science to Policy

Knowing that complex systems such as the climate, societies, or eco-systems can have critical thresholds, we obviously want to use those insights to manage for the best results. However, the reality of managing such complex systems is not straightforward. In the next chapter, I present examples of successful approaches to promoting "good" transitions and ways of preventing "bad" transitions. However, first I invite you to have a closer look at the treacherous road between science and policy. I do feel optimistic about the possibility of a paradigm shift in accepting and managing the chance of critical transitions. On the other hand, we should not be too naïve about the difficulties involved in moving from science to policy. Knowing the caveats is important if we want to make the best use of the science we can do.

A good starting point in our search for caveats is the fact that natural resources are usually of importance to several different groups in human societies (*stakeholders*). This is obvious for oceans and the climate system, but also smaller chunks of nature almost invariably serve different purposes valued by different groups of people. Lakes and streams, for instance, may be used by industries to get rid of wastewater but can also be used by swimmers who want clean water and by anglers who prefer certain kinds of fish. Almost invariably, there are conflicts of interest when it comes to decisions about the large and complex systems on which we depend.

The challenge is to find the best solution in such situations. To show the typical problems we encounter, I first describe an idealized way of designing socially fair environmental policy: striving at maximizing total utility, or *welfare*, obtained for society as a whole. Subsequently, I discuss why reality tends to differ widely from this theoretical optimum. Throughout, I will use the example of a lake to have a point of reference. However, the reasoning applies equally to the climate system, coral reefs, or any other system of natural resources that responds to human impacts. This view is based on work that I did with the economist William Brock and the sociologist Frances Westley; I follow the reasoning of an article the three of us published earlier, where more details can be found.[1]

16.1 Exploiting Nature in the Smartest Way

The approach we take as a start is that of an idealistic economist. A crucial step in finding the best solution for society as a whole is to express all interests in a common currency. In practice, this is often thought of as *money*; however, the concept can be expanded to include important nonmonetarian values, including elements such as happiness and ethical values to reflect something termed *welfare* or *utility*. After a brief explanation of how this philosophy works, I show how it can be linked to our models of the response of nature to stressors to reveal some eye-opening insights into the challenges of guiding societies' use of natural resources, especially if there are critical thresholds.

Stakeholders and Their Welfare

In the case of lakes, there may be many *stakeholders*, whose welfare is related to use of the ecosystem. Think for instance of farmers who allow nutrients from cattle dung and fertilizers to pollute the water in the catchment area of the lake. Reducing such pollution has a cost for the farmers. Thus, this use of the lake has an economical benefit for them. Similarly, households and industries that have their wastewater run into the lake avoid the cost cleaning it first. On the other hand, for anglers, swimmers, boaters, bird watchers, and waterfront land owners, the lake loses value if it becomes too polluted. Similarly, the income of hotels, campgrounds, and restaurants decreases if fewer recreational users are attracted by the lake, and drinking water companies that use the lakewater as a source will have to expend more effort if the water is polluted by cyanobacteria. Estimating how the welfare of each stakeholder changes with its use of the lake is not simple in practice. Also, one may argue whether maximization of the value for human use, instead of other ethical standards, should be the criterion to aim for. Nonetheless, the valuation approach is a large step forward as compared to simply leaving many obviously important values of ecosystems out of consideration in the decision process.

SHARING NATURAL RESOURCES IN THE BEST WAY

As a first step in a strategy to find the socially optimal use of nature, imagine the concept of a hypothetical *Rational Social Planner* (RASP). This RASP knows how the welfare of each stakeholder is related to its use of the lake that serves as our example and therefore should be able to decide what combination of uses would yield the highest welfare for society as a whole. However, to do this, the RASP needs to take into account how some uses of the system affect the value for others (for example, swimming is bad in murky water). Therefore, it is crucial that the RASP also knows how the system changes in response to its exploitation. It is the combination of the ecosystem response with the welfare functions that serves as a basis for the RASP to determine the integrated use that yields the highest welfare for society.

To illustrate the principle of maximizing welfare using knowledge of the constraints imposed by the functioning of the ecosystem, we start from simple response graphs, as you have seen earlier (figure 16.1). In these figures, the horizontal axes represent stress imposed, for instance, by human use such as nutrient loading. As argued, there is usually a clear economic benefit related to such use. We will call users that benefit from stressing the ecosystem *Affectors*. In contrast, we call users that benefit from the system but do not significantly affect the state of the ecosystem (such as swimmers) *Enjoyers*. The welfare of such Enjoyers thus depends on quality aspects such as water clarity

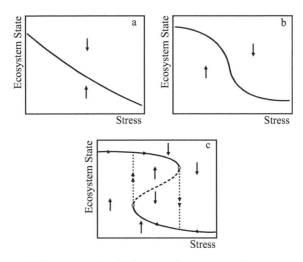

FIGURE 16.1. Different ways in which a complex system such as an ecosystem or the climate system may respond to stress resulting from exploitation by humans.

that we imagine as depicted on the vertical axes. In most cases, the ecosystem's value for Enjoyers will diminish with increasing exploitation by Affectors. Thus, in the graphs (figure 16.1), the low level of the system's state indicator at high exploitation will correspond to the lowest value for Enjoyers, and the welfare that Enjoyers can obtain from their use of the ecosystem will increase systematically with the level of the state indicator represented by the vertical axis. Obviously, many more groups of stakeholders exist in practice, and their interests may often be overlapping rather than strictly complementary, as in this Affectors–Enjoyers model. However, this distinction is useful to explain the basic idea.

If we assume that overall community welfare obtained from the ecosystem is simply that of the Affectors plus that of the Enjoyers, total welfare will increase along both axes used in the ecological response graphs (figure 16.2). If nature imposed no restrictions, the highest welfare could thus be obtained by combining maximum exploitation with a maximum value of the ecosystem's state indicator. However, the state is a function of the exploitation, and hence the ecosystem response limits the possible combinations of use by Affectors

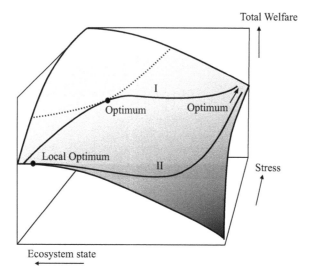

FIGURE 16.2. Graphical model showing how a theoretical society of Enjoyers and Affectors (see text) may obtain optimal welfare from the use of an ecosystem. The welfare of Enjoyers increases with the ecosystem state indicator, but the welfare of Affectors increases with the level of stress imposed on the system by their activity. Thus, total welfare will increase as indicated by the plane. The curves on the plane indicate how the ecosystem state responds to the imposed stress (as in figure 16.1). Optimum social welfare compatible with ecosystems dynamics is therefore obtained at the highest point of each curve. (From reference 1.)

and Enjoyers to points on the stable equilibrium lines in the response graphs (figure 16.1). Projection of these curves on the welfare plane (figure 16.2) shows in one image what combinations of use by Affectors and Enjoyers are possible and what their associated welfare is.

RECOGNIZING A BAD COMPROMISE

The highest point on each curve now represents the maximum overall welfare that a society can achieve. It seems obvious that we should want to aim at getting society as close as possible to such a maximum. However, the graphs reveal some tricky caveats. The simplest situation arises if there is a single optimum (figure 16.2a, curve I) at an intermediate stress level. In this case, a straightforward compromise between

Affectors and Enjoyers yields the highest overall welfare. However, in other situations, the results may be less intuitive. One complication arises when there are two alternative optimum points (figure 16.2a, curve II). Those represent biased situations with maximum welfare of either Affectors or Enjoyers. In this situation, a *compromise* (which is often the outcome of sociopolitical processes) may be the worst solution, as it represents a situation with low overall utility. Curve II in our example that results in this tricky situation represents the response of a *sensitive system*. Even low levels of stress result in a large deterioration of the state. The reason why a simple compromise yields low overall welfare in such situations is obvious if you think about it. Even a small stress level (yielding just a little profit for Affectors) produces a large loss for Enjoyers. Clearly, it is best to go for either of the two extremes rather than for the compromise now. If the system can be treated in separate spatial units (for example, if many lakes exist in an area), the obvious solution may be to assign some units entirely to Enjoyers and others entirely to Affectors.[2] However, this is not always possible, and in that case, it may be best to go for one of the two optima and compensate the group that loses in this situation.

MAXIMUM WELFARE AT THE EDGE OF THE CLIFF

The key question of course is how maximum welfare might be obtained from a system at risk of a critical transition to an alternative, less valuable state. Again, although we stick with the lake example, largely the same reasoning would be valid for many of the systems discussed throughout this book. Think of any system in which stress imposed by human activity may at some point tip the balance to trigger a critical transition into a state that is for some reason less beneficial to humanity. A clear lake may shift to a turbid state with toxic algal blooms because of pollution, a fishery may collapse because of overfishing, the thermo-haline circulation may stop heating Northern Europe beyond a critical point, or the Amazon forests may turn into dry savanna if too much forest is removed. What would be the optimum use of such systems? Typically, the maximum utility will be close to the threshold at which the system collapses (figure 16.3). The reason is that in such systems with a critical transition, stress typically has

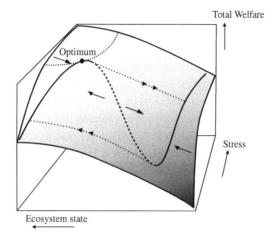

FIGURE 16.3. Optimum gains from the use of a system with alternative stable states can typically be obtained close to a situation in which a critical transition into the less valuable state is likely. This interpretation is analogous to figure 16.2. (From reference 1.)

little effect until stress has increased far enough to bring the system close to the border of collapse. Therefore, our Enjoyers will be well off until quite high levels of stress are imposed on the system.

The obvious implication is that aiming for the maximum welfare will tend to be a hazardous strategy. A slight "miscalculation" of our hypothetical RASP or some bad luck (for instance, an exceptionally hot year) may easily invoke a switch to the lower branch of the curve representing an alternative stable state with a low overall utility. In order to restore the system, the stress level now has to be reduced to quite low values (at the cost of a considerable further loss of total welfare) before a switch back occurs. In theory, the best strategy in such cases can be computed if the estimated costs and benefits in the long run are taken into account. Clearly, there is a trade-off between moving as close as possible to the optimum instant utility and preventing occasional costly collapse of the system. Such analyses have been done for hypothetical cases and show, not surprisingly, that it pays off to take a cautious attitude, especially if uncertainty about the distance to the threshold and the costs of collapse are high.[3]

16.2 Barriers to Good Solutions

Unfortunately, there is a huge gap between knowing what the best strategy would be and making it happen in reality. Why do we usually not deal with environmental resources in the best way from a social point of view? This question has answers on many levels. What follows is a bird's-eye view of what social scientists have revealed in this field.

The Invisible Hand Pushing Toward the Tragedy of the Commons

One of the early ideas about the problems we may run into regarding sharing natural resources is the "tragedy of the commons," explained by Hardin in an influential article in *Science*.[4] It is a reaction to the classic theory of the "invisible hand"—the idea that "an individual who intends only his own gain" is, as it were, "led by an invisible hand to promote . . . the public interest."[5] Indeed, there has been a broad tendency among economists to assume that the free market economy will automatically result in solutions that are best for society as a whole. In this view, no regulation is needed, as decisions reached individually will automatically be the best for society. Hardin's rebuttal to the invisible hand is borrowed from an 1833 pamphlet by an amateur mathematician named William Foster Lloyd and goes as follows: On communal pastures that open to all (commons), each herdsman will try to keep as many cattle as possible. This may work as long as tribal wars, poaching, and disease keep the numbers of humans and animals low. However, as those problems become less and the population grows, the picture of the tragedy of the commons unfolds. Each herdsman seeking to maximize his gain will ask, "What is the utility *to me* of adding one more animal to my herd?" This utility has a negative and a positive component. The positive aspect is that the herdsman receives all the proceeds from the sale of the additional animal. The negative component results from the additional overgrazing created by one more animal. However, since the effects of overgrazing are shared by all the herdsmen, the negative utility for any particular herdsman is only small. As a result, the hypothetical rational herdsman will conclude that the only sensible thing to do is to add another animal to his herd (and another, and another, . . .).

Since the same conclusion is reached by each herdsman sharing a commons, this will lead to a collapse of the resources, which is eventually a "tragedy" for all herdsman. As Hardin phrases it: "Ruin is the destination toward which all men rush, each pursuing his own best interest in a society that believes in the freedom of the commons."

THE SPILLOVER COST PROBLEM, OR WHY POLLUTING IS PROFITABLE

Clearly, Hardin's picture of the herdsmen it a bit naïve. Elaborate structures of cooperation, rules, and punishment to prevent havoc were developed early on in societies.[6] Nonetheless, it captures the essence of what may go wrong if common use of natural resources goes unregulated. Countless human activities affect our environment in ways that are undesirable from the point of view of large parts of society. In economic terms, such activities cause a "spillover cost" that is not taken into account by market prices.[7] For example, a motorcycle can give great pleasure to the one who rides it, but it generates irritation to those who have to listen to the noise. The market price of a motorcycle does not take into account the noise "costs" on other people. Hence from a social point of view, there is too much motorcycle riding under a free market system. In analogy to Hardin's tragedy of the commons, most environmental problems are examples of such uncompensated negative spillovers. In terms of our lake model, in an unregulated situation, Affectors benefit from their activities, while the costs resulting from a deteriorated ecosystem state are carried by the Enjoyers. In the common situation that Affectors are also partly Enjoyers of the same ecosystem, the costs of the activities may be considered to be turned to the community as a whole, whereas the profit from the affecting activity goes exclusively to the Affectors. This bias is the core of many environmental problems.

THE COLLECTIVE ACTION PROBLEM, OR WHY WE CAN'T DO MUCH ABOUT IT

The first step in the direction of a fairer situation is to mobilize the forces of Enjoyers in order to press for regulation. However, this often does not work, for a reason known as the *collective action problem.* Game theoretical models that address collective action problems suggest that group effort will be systematically lower in large diffuse groups than in small concentrated groups. This is because of low *perceived effectiveness* and *noticeability.*[8] The larger a group, the more anonymous each member will tend to feel. Hence, self-interest may lead each individual in a large group to shirk the duty of contribution of a "fair" share of the group effort. Obviously, the drop of individual effort with group size depends on how effective the group is in making each member feel "noticeable" so that that member pulls his or her own weight in the joint effort.[9]

The graphical models that show how the welfare of society could be maximized can be modified to produce graphs that show the expected outcome of political pressure. The change of focus is that rather than seeking the social welfare optimum, the authority that regulates the system is responding to political pressure. As argued earlier, political pressure depends on the interest at stake (that is, the welfare, as shown in figure 16.3) but also on the effectiveness of the interest group to mobilize forces. We can obtain a graph that represents the political force that can be applied by Affectors and Enjoyers to obtain a certain utility from the ecosystem by multiplying that utility with a factor that represents the ability of the group to mobilize forces (figure 16.4). In a situation in which the Enjoyers are a more coherent and concentrated group than the Affectors, the Enjoyers' political power will be relatively strong. In the case of a system with alternative stable states, this will tend to lead to an equilibrium that is on a relatively safe part of the "good" branch of the equilibrium curve (figure 16.4a). The resilience of this situation is relatively high. However, usually Affectors are better organized than Enjoyers, who may often be a large but diffuse group. As a result, the political power of Affectors is often disproportionally high, resulting in a situation in which there is no local optimum representing a power equilibrium on the "good" branch of the curve (figure 16.4b). Instead, the political pressure will drive society further and further up along the branch with low Enjoyer value, because of the high pressure produced for even slight gains of Affector utility.

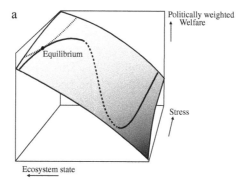

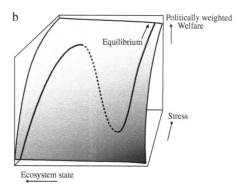

FIGURE 16.4. Differences in efficiency at mobilizing political pressure distort the process of optimization of social welfare, depicted in figure 16.3. The system will tend to an equilibrium in which political pressure from different interest groups is in balance. If Enjoyers are more efficient (panel a), that equilibrium will be on a more resilient part of the branch representing the desired ecosystem state. However, typically Affectors are the more efficient group at mustering political pressure (see text) resulting in a situation (panel b) where the system tends to increasing stress on the ecosystem, even after it has collapsed to the lower branch of the curve. (From reference 10.)

The Delayed Response Problem, or Why We Are Always Late

The collective action issue is not the only problem when it comes to our failure to regulate the use of complex systems such as the climate or ecosystems in a good way. At any instant in time, many activities that cause environmental deterioration are not regulated at all. This is inevitable because of the continuous development of new activities resulting in new environmental problems. Clearly, societies that are slower in detecting and regulating new problems will carry a higher overall cost. Therefore, it is worth looking into this problem a bit more.

Part of the explanation for slowness in response may be the difficulty of detecting a new problem. Much as in an immune system, detection of a new problem will depend on past experience.[11] If a problem is unlike anything that has been encountered in the past, detection may take a long time. An example of a problem that has existed for a long time without being detected is that of so-called *endocrine disrupters*. The fact that numerous chemical substances of widely varying origin may disrupt the endocrine hormone systems of animals and humans was simply not known for a long time. Now, the potential of chemicals released into the environment to disrupt the chemical communication between organs, tissues, and cells in organisms has become a major cause for concern[12] and a strategy for detecting new instances is being worked out.[13]

Unfortunately, detection of a problem by scientists does not guarantee a quick regulation. A first delay on the way to regulation is due to the inertia in public opinion discussed in some detail earlier (section 12.4). A shift to recognition of a problem may occur only after a long period, as opinion remains locked in a passive attitude. As argued, this effect is stronger if problems are more complex so that it is difficult to make up one's mind on the basis of one's own observations. In that situation, an individual depends more on the opinions of peers and authorities. The stronger peer effect increases the hysteresis and therefore the inertia, whereas the dependency on authorities and opinion leaders increases the risk of manipulation. In light of the power bias discussed in the preceding section, this implies a larger scope for powerful groups of stakeholders to downplay the importance of a problem successfully, simply by hiring good "sense makers" and getting their worldview well exposed.

Even if a problem becomes widely recognized, it can take a long time before effective regulation occurs. How long this phase is can vary significantly from one situation to another. Obviously, if the proposed regulation of a problem requires giving up some wealth by a concentrated powerful group, this group may effectively delay or block the process here.[8] A second variable that can have a large impact on the time it takes to move from recognition to regulation is the distribution of decision making power in society. In highly centralized/more authoritarian decision making structures, once the central authority is convinced of the need to change, the system can react more quickly and with tight coordination. In a decentralized system, where decision making authority is equally distributed across all parties, change demands a negotiated agreement to coordinate actions.[14] Although such solutions may be relatively sustainable,[15] they can take a long time. Clearly, it is most difficult to regulate a problem if there is no central decision making authority and if the distribution of benefits from the current situation is unequal. In these circumstances, negotiations are necessary, and stonewalling is likely to occur on the part of privileged actors. Many environmental issues that are truly "global," such as global warming, may fall into this category. While certainty about the problem and the need for action is fairly high, it is unclear whether effective action is possible at all, as illustrated by the problems in making the Kyoto Accord work.

Obviously, the problem of slow response is especially critical in situations in which an ecosystem or the climate system is about to cross a critical threshold for a catastrophic shift. If the problem is unlike something we have seen before, which may be the case for rapid climate change scenarios, first detection may already be too late to allow any prevention of the switch even if further delays to regulation are avoided. Also, as the perceived seriousness of the problem will seem small until the irreversible switch occurs, the shift to general recognition of the problem will tend to occur too late. Certainly, sophisticated integrative solutions are not easily reached under such situations of urgency. On the contrary, as argued earlier (section 12.6), stress tends to result in rigidity rather than innovation and flexibility. In any case, delay in decision making is likely to be long in global problems as no central authority exists and measures will be most costly to the most powerful countries. All this suggests that the diagnosis of delay mechanisms may be particularly relevant for designing policy strategies for preventing sudden irreversible shifts resulting from global change such as a collapse of the thermo-haline circulation.

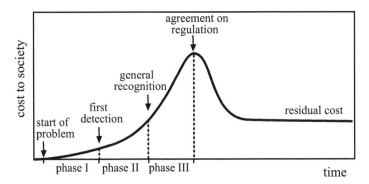

FIGURE 16.5. The costs to society of a new activity that causes a *spillover* problem are initially very small but will grow as the intensity of this activity increases. There may be a long time lag until the moment of regulation of the problem in which three phases can be distinguished: (I) a period in which the problem goes undetected altogether, (II) a period in which general recognition of the problem is lacking, and (III) a delay before the onset of actual regulation. Since eventual regulation is typically not reflecting a socially optimal solution, there remains a residual cost to society. (From reference 10.)

16.3 Synthesis

In conclusion, while humans are innovative in the use of natural resources and finding solutions to various problems, there are also some fundamental reasons why human societies tend not to get the most out of natural resources when those become limiting or act appropriately if resources approach the verge of collapse. There is a tendency to activities that benefit a subgroup of society, but at the cost of the condition of ecosystems and environmental quality. Regulation of such problems may not happen at all. Also, if it happens, it typically does not reflect the best solution for society as a whole, and may come only after a long time because of late detection by scientists, inertia in the public agenda, and delay or gridlock in regulation (figure 16.5). Entrepeneurship naturally drives a permanent search for unregulated situations where it is still possible to "communize the cost and privatize the profit." This causes new environmental problems to arise continuously (figure 16.6). As a result, society may carry a high cost of environmental problems at any instant in time due not only to

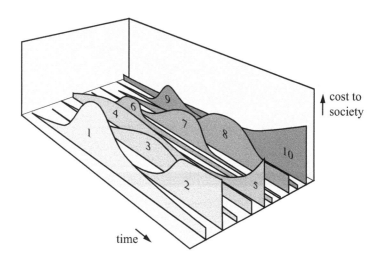

FIGURE 16.6. New problems continuously arise, and the total costs of all these problems to society depend on the ability to recognize and regulate problems well but also rapidly. Some problems may be regulated relatively quickly (for example, problem 6) and be eliminated almost entirely (for example, problems 6 and 7), whereas others grow unregulated for a long time (for example, problem 10) or correspond to irreversible switches that cannot be solved (for example, problem 2). The area below the curve of a specific problem represents its cumulative cost to society starting from the moment of its introduction. The sum of the costs of all individual problems at one instant of time is the total environmental spillover burden to society at that moment. The grand total of environmental spillover costs carried by society is the sum of the areas under all the curves. (From reference 10.)

suboptimal regulation, but also to failure to recognize new problems, or slowness in regulating them. The risk of such failures is particularly high if the natural system has a threshold at which a shift into an alternative basin of attraction may occur. This is because obvious signs of deterioration remain minor until the breakpoint is reached. The costs of such collapse can be very high, as restoration of the preferred state may take a long time, if it is possible at all.

CHAPTER 17

New Approaches to Managing Change

Now that we have examined the caveats of going from science to policy, let us move forward and look at the situation from the positive, practical side. What can we do, and how? First, we will look at how good transitions can be promoted and how bad ones might be prevented. Clearly, it is not a straightforward matter to agree whether a particular transition is good or bad. Some transitions are widely considered bad.For instance, we do not like the collapse of cod. Also, we would not want the thermo-haline circulation to shut off. The resulting dramatic temperature drop in northern Europe and the northeastern United States would be simply too disruptive. On the other hand, we like people to escape the poverty trap, and we appreciate a lake shifting from a murky to a clear state. However, as discussed in the previous chapter, "good" is often not good in all aspects and for all stakeholders. Similarly, "bad" is often not bad for all either. Nonetheless in this chapter, we return for the sake of clarity to a simple dichotomous distinction between good and bad transitions, and contemplate the corresponding questions of how we can promote the good transitions and prevent the bad ones. I first highlight some examples of both categories and then wrap up by summarizing what can be done in terms of managing resilience, building adaptive capacity, and triggering transitions.

17.1 Promoting Good Transitions

Finding smart ways to promote a self-propagating shift from a deteriorated state to a good state is perhaps the most rewarding part of the science of critical transitions. Doing this can be quite easy once you find the Achilles' heel of the system. In its most beautiful form, the process goes like this: Find out how to reduce the resilience of the bad state first, and then flip out of the bad state with little effort. I will just let a few examples speak for themselves.

Biomanipulation of Shallow Lakes

The theory of alternative stable states has become a cornerstone of restoration strategies for shallow lakes. The pristine state of such lakes is mostly one of clear water with extensive submerged plant beds hosting a rich community of invertebrates, fish, and birds. However, an overdose of nutrients from agriculture and wastewater has turned the water of many lakes into a greenish soup with a poor biodiversity and frequent blooms of toxic cyanobacteria. The stability properties of shallow lakes were discussed in some detail earlier (section 7.1). The essence is that the clear and the turbid situation represent alternative stable states over a range of nutrient loading. This was discovered by ecologists involved in restoration of such turbid lakes. As I was one of those ecologists, what follows is my version of the inside story of this discovery. It turned out that upon reduction of the nutrient loading to levels at which the lakes used to be clear, many of the turbid shallow lakes remained turbid. An observation by Czech biologists working on fish production ponds set us on track.[1] They had found that ponds with few or no fish were consistently more transparent than ponds with high fish biomass. Also, they found a systematic effect of fish on the zooplankton community that was later confirmed by John Brooks and Stanley Dodson in a now classic paper in *Science*.[2] Faced with the problem of lakes that would not return to their former clear state, we decided to see what would happen if we took fish out of a lake, an approach that became known as *biomanipulation*. We started with small experiments in enclosures and fish ponds, and after the results turned out well, we went to try it in small lakes.

Figure 17.1. After pumping out part of the water, fish is removed from a small swimming lake in The Netherlands to induce a transition from a turbid to a clear state.

One of the first lakes we tried was a tiny lake close to a village in Holland (figure 17.1). As the name of the lake, Zwemlust ("swimming joy"), suggests, it had always been used intensively by the people from the nearby village for swimming. It even had some commodities turning it into a kind of natural outdoor swimming pool. Over the years, however, the lake went the same way as most lakes in the densely populated countryside of The Netherlands and turned into a murky greenish pool. The phytoplankton concentration in the lake was very high, and blooms of toxic blue-green algae frequently turned the lake bright green. After several unsuccessful attempts to improve the water quality, it was decided to manipulate the fish stock. This measure had spectacular results.[3] In March 1987, the water was pumped out of the lake to facilitate complete removal of the fish. It appeared that 1,000 kilograms per hectare of fish had been present. Seepage from the nearby river refilled the lake in three days, and a small fish stock was introduced together with waterfleas and some aquatic plants. Shortly after the refill, there was an algal bloom, but soon large waterfleas became abundant and grazed down the algal biomass to a mere 2% of the premanipulation values, and the water became crystal clear and stayed so for years.

The news spread rapidly, and the experiment was followed by many more. Eventually, we scaled the approach up to trawl nets of 3 kilometers in length through some of our largest lakes. Invariably, the lakes turned clear provided that more than 75% of the fish had been removed.[4] Fish populations quickly recovered in most lakes but stabilized into a different community configuration. Most importantly, submerged vegetation recovered in the lakes, stabilizing the clear state for years to come. Similar work was done in parallel in Denmark and Britain, and the approach became a well-accepted part of the toolbox for lake restoration.[5] Interestingly, in each of the countries, a single key person may be identified as instrumental in promoting the new approach. Harry Hosper (The Netherlands), Erik Jeppesen (Denmark), and Brian Moss (Great Britain) are all scientists, but also charismatic "sense makers" who phrased the problems and the novel solutions in ways conducive to action for their contacts in the policy and water management world.

Twenty years after the first experiments, it turns out that the long-term response of the lakes has been diverse. Some lakes turned turbid again soon after the experimental perturbation, others stayed in the clear state for as long as we know, while a large intermediate group stayed clear for years, but eventually slipped back into the turbid state.[6] All this fits well into the theory, suggesting that the stability of the clear state depends on the nutrient status of lakes (figure 17.2). The long transient response of some lakes has been explained as an effect of a *ghost equilibrium*. Even if the clear state is not strictly stable, its ghost may practically paralyze the system, making change very slow in its vicinity (section 5.1). From a practical perspective, such long transients are quite interesting, as a relatively cheap shock therapy such as biomanipulation can imply a long-lasting improvement of the ecosystem state.

Using El Niño for Dry Ecosystem Restoration

A parallel to the biomanipulation of lakes has been developed for the restoration of forests in semiarid regions (figure 17.3). Many of the semiarid landscapes around the world were originally covered by extensive shrublands and forest. However, the original vegetation is now

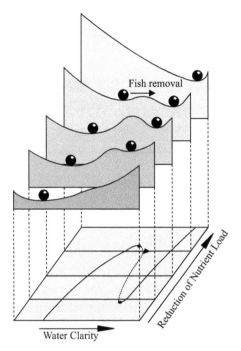

FIGURE 17.2. Fish removal can flip shallow lakes from a turbid to a clear state. However, long-term stability of the turbid state depends on whether nutrient levels have been reduced far enough.

reduced to a few remnant patches in most countries. After wood extraction, some areas have been put to agricultural use, but often loss of fertile soils allowed this practice to continue only briefly. A poorly vegetated state in which shrubs are scattered in a matrix of sparse herbaceous vegetation is a common endpoint. The economic value of such landscapes where goats are virtually the only product is low.

Efforts to restore original dry forests have been unsuccessful in most of these areas. Rabbits and goats (both introduced by humans throughout the world) play a key role in prohibiting the vegetation from recovering. Another barrier to the recovery of the original forests is drought. In the absence of shade from adult trees, seedlings of the typical species that formed the forests survive only during exceptionally wet years. Precipitation may be an order of magnitude more during El Niño years in some regions and often leads to a spectacular

FIGURE 17.3. The central valley of Chile was once covered by dense woodlands. Now, only isolated patches of trees remain, and soils are largely degraded. The savanna-like system in the foreground is maintained by introduced grazers propagating the introduced *espinal* shrubs.

pulse of plant growth. However, this productivity increase is invariably followed by an increase in rapidly reproducing herbivores that quickly hammer down biomass to previous levels.

Those dynamics brought the ecologist Milena Holmgren to the idea to trigger forest recovery by excluding herbivores from designated areas precisely in tune with El Niño rains.[7] The occurrence of an El Niño phase has become increasingly predictable, to the extent that rainy periods can be foreseen months in advance. This gives sufficient time to act and exclude grazers and disperse seeds at the moment at which the resilience of the degraded state is lowest. Once the seedlings have grown sufficiently, they become quickly less vulnerable to both drought and herbivory. Moreover, once a canopy has formed, this allows subsequent natural regeneration of the forest, as the water conditions are much better for seedlings in the relatively cool and moist understory.

Note the similarity but also the subtle difference to the practice of biomanipulation in lakes. Just as in lakes, the abundance of a key functional group of species that keeps the system in the unwanted state is temporarily pushed down. Also in both cases, it is considered important to apply this disturbance in a situation in which the resilience of the unwanted state is low relative to the resilience of the preferred state. However, in the case of lakes, the resilience is usually altered first by reducing the nutrient loading, whereas in the case of forest restoration, the idea is to wait for a moment at which such a window of opportunity with altered resilience naturally occurs (figure 17.4). Clearly, this is an elegant way of making use of the natural swings in the forces of nature.

Microcredits to Escape the Poverty Trap

Poverty is often considered a self-reinforcing stable state, the *poverty trap* from which it is difficult to escape (see also section 2.2). Clearly, this is one of the major problems facing humanity. There are about three billion poor people, and roughly a quarter of the global population has to survive on less than a dollar a day. State collapse and wars hit poor regions most often,[8] and deep poverty is considered one of the main impediments to sustainable peace in many places around the world. No wonder that in 2006, the Nobel Peace Prize went to an economist who found a way that has allowed many families to escape from the poverty trap. His solution, *microcredit*, had already been celebrated the year before by the United Nations Development Program (UNDP), declaring 2005 the international year of microcredit. Their logo (figure 17.5) aptly demonstrates the idea of escaping from a trap.

The trap aspect that microcredit solves is the lack of access to loans by poor people. Banks traditionally considered the risk that the money would not be returned to be too great. Around 1974, Muhammad Yunus, the Bangladeshi economist who received the Nobel Peace Price, determined that this idea was wrong. In an interview with the *New York Times* (October 14, 2006), he recounts his eye-opening experience. He gave $27 to be shared among a group of 42 Bangladeshi people who asked for help, including a woman who made bamboo furniture, which she sold to support herself and her family. They soon made money on his money and repaid him as if his gift had been a loan. "If you can make so many people happy with such a small amount of money," Yunus recalled, "why shouldn't you do more of it?"

It turned out that for many (but not all) poor people, it was possible to escape from the poverty trap once they had just the little bit of

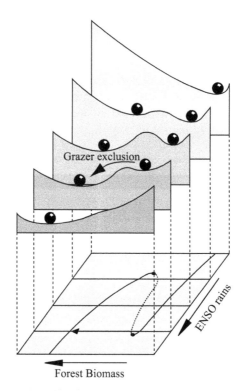

FIGURE 17.4. Grazer exclosure in rainy years may flip degraded drylands to a forested state. Note that ENSO enhancement of rainfall alone may be insufficient to induce the change without grazer exclosure (all landscapes except the front one). Similarly, grazer exclosure may not help if rainfall is not elevated beyond the base level (last landscape).

money needed to invest in a tiny business of their own and start making more out of it. They often did not need more loans—just a one-time *microloan* that they paid back soon. In terms of our theoretical framework (section 14.2), the system was shown to shift to a stable contrasting state upon a single perturbation, implying that the poverty trap was experimentally proven to be an alternative attractor to a more wealthy state. Of course, the theory also tells us that alternative stable states exist only over a limited range of conditions. No wonder then that microcredit has been found to be a solution in only a subset of cases. Some environments where loans are made are simply too

FIGURE 17.5. The UNDP Year of Microcredit Logo illustrates the idea that microcredit can promote a critical transition out of the poverty trap.

poor to support small-business growth. Also the Nobel Prize Committee realized the limitations. As they state in their press release: "Yunus' long-term vision is to eliminate poverty in the world. That vision cannot be realized by means of micro-credit alone."

Nonetheless, the impact of microcredit has been undeniable. In fact, the idea itself has a longer history. However, the credit goes to Yunus especially, because he pursued it further in a remarkably successful way. A few years after his eye-opening experience, he founded the Grameen Bank to make loans to the poor people in his country. Since then, the Grameen Bank has issued more than $5 billion in loans to several million borrowers. To ensure repayment, the bank uses a system of "solidarity groups": small informal groups, nearly all of them exclusively female, that meet weekly in their villages to conduct business with representatives of the bank, and who support each other's efforts at economic self-advancement. The success of the Grameen Bank did not go unnoticed. In November 1997, delegates from 100 countries gathered at a Microcredit Summit in Washington, with the goal of reaching 100 million of the world's poorest families by the year 2005. Support for these goals has come from prominent world leaders and major financial institutions. The World Bank estimates that there are now more than 7,000 microfinance institutions, serving some 16 million poor people in developing countries.

Why would so many institutions be so interested in the idea, spreading the practice so rapidly over the world? Analyses published in journals such as *The Economist* consistently show that the interest rates charged on microcredit are usually much higher than those on normal loans. This might explain the enthusiasm of big financial institutes to invest in this business. Microloans simply turn a "market failure" into a market success. This is not to criticize the work of Yunus, who promoted his insight into a way to effectively trigger a critical transition in the lives of so many impoverished people.

To quote the press release of the Nobel Committee once more:

> Loans to poor people without any financial security had appeared to be an impossible idea. From modest beginnings three decades ago, Yunus has, first and foremost through Grameen Bank, developed micro-credit into an ever more important instrument in the struggle against poverty. Grameen Bank has been a source of ideas and models for the many institutions in the field of micro-credit that have sprung up around the world.

17.2 Preventing Bad Transitions

Successful examples have played a key role in the rapid spread of some innovative ways to trigger "good" transitions. Unfortunately, the flip side of the role of examples implies a fundamental problem when it comes to the prevention of unwanted transitions. What if you do not have a successful example? How can you show that you prevented something bad from happening? Or worse, how can you convince people to invest in preventing something bad from happening of which no obvious signs are visible? Earlier, I discussed some of the deeper socioeconomic aspects involved in this problem (chapter 16). Here, I give a few examples of how attempts have been made to prevent a critical transition from a good to a bad state.

Preventing a Transition to a Turbid State in Lake Veluwe

Not surprisingly, lakes are among the rare examples of systems where management policies have explicitly addressed the issue of preventing an unwanted critical transition. In a general sense, a theoretical study of a hypothetical lake has addressed the question of how one may best deal with a poorly quantified risk of collapse from a clear into a turbid state, given assumed economic implications and costs of management.[9] However, some lakes are sufficiently well understood to move beyond such abstract studies. An example is the large and shallow Dutch lake Veluwe. This lake is heavily used for recreation and has been studied intensively for decades. Its history includes transitions from the clear to the turbid state and back, and the hysteresis in its response to changes in nutrient loading as well as the processes involved have been closely scrutinized.[10] The lake is now clear, but a renewed increase in nutrient load is expected. Also, plans to increase the depth of parts of the lake and remove aquatic vegetation jeopardize the resilience of the clear state. The water management authorities are aware of the risk that the lake would fall back into the turbid state dominated by cyanobacteria and asked the scientists to estimate the probability that such a collapse would happen under different scenarios. A stochastic model is now used to estimate such risks[11] and provides a basis for rational decision making. For instance, the probability of exceeding critical values for key parameters, such as the total phosphorus concentration and water transparency, taking into account annual variation in meteorological conditions and random noise in ecosystem relationships, is used as a criterion for decision making on phosphorus emission reduction measures.

Building the Resilience of the Great Barrier Reef

The Great Barrier Reef is the largest reef ecosystem in the world. Consisting of about 2,900 individual reefs, it extends for 2,000 kilometers along the Australian East Coast. It is of tremendous aesthetic value, and tourism on the reefs contributes substantially to the economy. Although at first glance, this impressive ecosystem may seem close to virgin, it is in reality showing serious symptoms of ecological change.[12] Export fisheries that flourished after European colonization (for example, for sea cucumbers, pearl shell, and turtles) have collapsed or are no longer commercially viable, and large-scale outbreaks of crown-of-thorns starfish have reduced coral cover. A particularly important wake-up call came when major bleaching events from climate change struck the Great Barrier Reef in 1998 and 2002, causing damage to more than 600 individual reefs. It was felt that a combination of threats ranging from runoff of sediment and nutrients to exploitation of megafauna and fish stocks had reduced the resilience of the reefs to overcome shocks.[13] The collapse of the Caribbean reefs into an algal-covered state had been a clear example of how overfishing and other stresses can make such systems vulnerable to large-scale transitions that seem difficult to reverse. Since, in the face of ongoing climatic change, the threat from bleaching events and ocean acidification would likely increase, it was decided that the ecosystem should be managed for resilience. Following a series of seascape experiments, a bold large-scale management plan has been implemented comprising a series of measures including closure of 33% of the reefs from fishing.[12]

PREVENTING THE COLLAPSE OF COD

The management of commercial fish stocks is another obvious example where there is an effort to find ways of preventing collapse into an unwanted state. For lakes and reefs, cases of collapse have been documented, and this helps to bring the issue to the political agenda. Unfortunately, despite the efforts of many fisheries biologists worldwide, the mechanisms that drive the dynamics of fish stocks are still poorly understood. Fisheries clearly have a huge effect on many stocks, but shifts in marine circulation patterns and other climatic effects may have large effects too. Teasing such mechanisms apart for these huge and open systems is simply a daunting task compared to what is needed to unravel the forces driving the dynamics of a lake. One of the eye-catchers when it comes to fisheries is the collapse of cod off the coast of Newfoundland (section 10.1). One could argue that this was one of the more intensely studied and managed fish stocks on the Earth. So why is it that the dramatic and apparently irreversible collapse could not be foreseen and prevented? Most likely, the system was still poorly understood, and the models used by the fisheries scientists were therefore inadequate.[14] However, in addition to uncertainty on the part of science, problems in the process of approving and enforcing management policies are typically huge in marine fisheries.[15] As I write this book, European Union fisheries scientists have advised closure of North Sea cod fisheries entirely, as the risk of a complete collapse seems real. Politicians from the different countries that fish in the area are reluctant to follow this advice, and we can only hope that on the day you read this book, cod is doing well again in the North Sea.

Preventing Critical Transitions in the Climate

Preventing the collapse of fisheries is difficult in view of the large scientific uncertainties and the difficulties involved in designing and enforcing policies when many stakeholders in different countries are involved. Clearly, preventing critical transitions in the climate system will be even more difficult for the same reasons. Models and reconstructed dynamics suggest alternative attractors in the Earth system. However, much uncertainty remains. Unlike in the case of lakes or fisheries, we have no clear and well-documented examples of critical transitions in recent times. Closest in time, the Sahara was born rather abruptly 5,000 years ago. Perhaps better known is the collapse of the thermo-haline circulation, which has happened several times (most recently, around 9,000 years ago). The movie *The Day after Tomorrow* dramatized the possible effects, but with little realism. Probably, resilience of the current state is much higher than during the last glacial maximum when a collapse occurred. However, uncertainty in our understanding of this part of the climate system remains very high,[16] and the same holds for other areas of potential critical transitions such as the Arctic and the Amazon basin (section 11.1).

Our management options for preventing potential rapid climatic transitions are largely restricted to managing atmospheric CO_2. This is tremendously difficult, as it requires globally coordinated action. As recognized by the Nobel Prize Committee, efforts from the Intergovernmental Panel on Climate Change (IPCC) and also from former U.S. Vice President Al Gore and other key players have eventually triggered a major shift in the public attitude toward seeing climate change as a human-driven threat. Signs of change that are readily visible to anyone have undoubtedly contributed to this shift in attitude. Nonetheless, it remains to be seen whether a globally concerted action to reduce net greenhouse-gas emissions can be invoked that would be substantial enough to prevent major climatic change, including critical transitions in sensitive subsystems such as the Arctic and the Amazon basin.

17.3 Synthesis

Perhaps the most striking aspect of the examples discussed here is the ease with which approaches to triggering "good" transitions spread. They may rightly be considered paradigm shifts, changing the way people see a particular problem in a fundamental way. Clearly, convincing examples of how spectacular change can be brought about with little effort may be essential. The cases of lake restoration and microcredit illustrate that. In the case of microcredit, the profit that microloans can bring to the institutions that provide the credit may be another dimension of the explanation. However, the aspect of "doing something good" is an undeniable part of its appeal too.

The strategy of building resilience to prevent a "bad" transition is used in places like the Great Barrier Reef and Lake Veluwe, but the idea is clearly less contagious than that of triggering good transitions. Rationally, the argument for working on resilience seems straightforward. Resilience is usually more easily managed than the stochastic events that may trigger transitions: it is difficult, but not impossible to work on trends in global warming, pollution, land cover, poverty, and exploitation. In contrast, we will never be able to avoid an incidental drought, hurricane, biological invasion, terrorist action, or evil leader. Perhaps one problem in promoting acceptance of this view is that intuition simply points the other way. Intuition suggests that major shifts in nature or society have distinct causes such as a drought, a hurricane, an earthquake, an evil leader, or a meteor rather than an explanation in terms of loss of resilience. Clearly, intuition is not necessarily always our best guide. It also tells us that the world is flat. But, unfortunately, it is more difficult to show that complex systems can become fragile in an invisible way as a result of gradual trends in climate, pollution, land cover, poverty, or exploitation pressure than to demonstrate that the Earth is round. The key role of examples in convincing people is simply problematic when it comes to preventing a collapse into an unwanted state. It is difficult to convince policy makers to invest in preventing something bad from happening if no obvious signals of upcoming trouble are visible. Also, in hindsight, it is difficult to show that the investment actually prevented a bad transition. On the other hand, the power of examples promotes thinking about resilience when it comes to triggering good transitions. In lakes, nutrient levels are reduced in order to reduce resilience of the turbid state and increase that of the clear one. Similarly, microloans are known to work only if overall poverty is not so deep that it would prevent the wealthier state from being stable and make resilience of the poverty trap too large.

A nice application of the theory of critical transitions is the idea to tune management to natural variation in resilience. This is relevant for promoting good transitions as well as for preventing bad transitions. Natural swings may open windows of opportunity to induce a transition out of an unwanted state. For instance, a rainy El Niño year may be a window of opportunity for forest restoration, and a year with low water levels may make it easier to push a shallow lake to the clear state. On the other hand, rangeland managers in Australia anticipate droughts that hit the continent during El Niño years by reducing livestock to prevent potentially irreversible degradation of the land that may result from overgrazing in such years.[17]

Clearly, managing resilience in large systems such as the climate, societies, or extensive marine systems remains a challenge. Nonetheless, the success of managing critical transitions on smaller scales suggests that much can be gained by applying these ideas to a wider range of complex systems.

CHAPTER 18
Prospects

18.1 The Delicate Issue of the Burden of Proof

While the idea of tipping points sounds compelling, there is an obvious tension between the precision of mathematical theory and the complexity of mechanisms that determine stability and transitions in real complex systems. It is not easy to strike a good balance between realism and precision. Strictly speaking, equilibria and bifurcations obviously do not exist in the real world. Nonetheless, the "mirror world of math" reveals some fundamental features of reality that would otherwise be difficult to comprehend. In the past, concepts such as chaos and catastrophe theory have sometimes been applied all too loosely. This does not help much if we want to unravel what drives change in complex systems. On the other hand, discarding attempts to link theory to real systems too lightly as imprecise may not be a good idea either.

At first glance, it makes sense to accept a new hypothesis only if there is good evidence. However, in the case of critical transitions, the question is what our default assumption should really be. In light of the theory and evidence summarized in this book, it would be remarkable if ecosystems, the climate, and societies did not have tipping points. This implies that one could argue that instead of requiring the proof that a system has tipping points, it might be better to request a proof that the system has only one stable state. This implies that one should show, for instance, that from any initial condition, the system eventually settles to the same regime. Obviously, this is very difficult in practice. However, one may question where the burden of proof should be. Assuming that tipping points are absent when they are actually present may lead to dangerous false assumptions, such as the idea that pollution effects can be easily reversed, that the climate will change smoothly, and that a harvested population will not collapse.

18.2 Toward a Practical Science of Critical Transitions

In this view, a general precautionary principle makes sense, but it would be much better if we could predict tipping points. The search for generic early warning signals is an exciting development, as it may help to assess the risk of a transition even if the mechanisms are unknown. On the other hand, if we also want to find smart ways to manage critical transitions, an understanding of the governing mechanisms is obviously a must. There is no silver-bullet approach to obtain such insight. The strongest science is traditionally built on controlled replicated experiments. However, for studying the stability properties of ecosystems, societies, and the Earth system, such experiments are impossible on meaningful scales, and we necessarily have to rely on interpretation of natural experiments and models that integrate our insight into key processes.

It may seem a daunting task to unravel the functioning of vast and complex systems such as societies, ecosystems, or the climate. Nonetheless, looking back on this book, I feel there is good reason for optimism. Science has already inspired policies for building resilience of large systems such as the Great Barrier Reef, and we are finding smart ways to invoke critical transitions out of undesirable situations such as poverty traps, turbid lakes, or degraded drylands. Perhaps one explanation for the success of unraveling the causes of critical transitions even in large complex systems is the fact that such shifts are often propelled by relatively simple mechanisms. Such governing mechanisms can be understood more easily than the myriad of details. There will always be large uncertainties in what we know about complex systems, not to speak of the problem that we do not know what we do not know. A pessimist might say that our knowledge is hopelessly fragmentary. However, the good news is that we really do not need to know everything when it comes to understanding tipping points. The challenge, of course, is to have an eye open to the things that matter and to avoid being distracted by the details that do not, even if we happen to know a lot about those irrelevant details. The natural course of most science is to elaborate on things we know, creating densely explored patches of splendid insight in a vast landscape of poorly known elements. Examples of unraveled tipping points often involve the connection of such isolated patches of knowledge bridging unknown territories.

Perhaps you are seeing critical transitions everywhere after reading this book. While the divide between fascination and obsession can be delicate at times, a search image for the architecture of tipping points may not be so bad. Certainly, gradual change is the rule and critical transitions are the exception. However, in view of the formidable implications of some of these transitions, they certainly are an exception that is worthy of special attention.

Appendix

This appendix gives mathematical models for a selection of the phenomena that are discussed in this book. Rather than trying to be mathematically rigorous, I have attempted to make the discussion accessible to most readers, explaining formulations starting from the basics. As a result, this appendix can be seen as a kind of mini-guide to modeling dynamical systems. Readers already familiar with modeling can use this as a reference for basic formulations related to the theory explained in the text. Different software packages can be used to play around with the model. The package GRIND for MATLAB is versatile and convenient for the types of analyses referred to in this book. It is freely available at *http://www.aew.wur.nl/UK/GRIND/.*

A.1 Logistic Growth

Probably the best-known growth model is the *logistic equation.* It describes population increase (dA/dt) as a function of population density (A) and two parameters:

$$\frac{dA}{dt} = rA\left(1 - \frac{A}{K}\right).$$

The properties of this simple model are illustrated in figure A.1. The relative growth rate ($dA/dt/A$) is highest (r) when the population density (A) is very low relative to the carrying capacity of the environment (K).

In that case, the competition term $(1 - A/K)$ is close to one. The logistic equation assumes competition to cause a linear decrease of the relative (*per capita*) growth rate $(dA/dt/A)$ with the population density (figure A.1a). The result is that the overall productivity (dA/dt) has a maximum at half the carrying capacity (figure A.1b). When the population density is too low, the productivity is limited by the amount of reproducing individuals (A); when the population approaches the carrying capacity $(A = K)$, productivity tends to zero because of competition. Note that the logistic equation predicts that optimal harvest can be obtained by maintaining the population at a density of half its carrying capacity. Plotted against time, the density of a logistically growing population follows a sigmoidal curve, with the steepest growth occurring at half the carrying capacity (figure A.1c).

Its simplicity makes the logistic equation very attractive for use in simple models, but obviously, the model is a crude simplification. A strictly linear decrease of per capita growth with density seems unlikely to be found in any real-life situation. Also, in a simple model like the logistic equation, many underlying regulatory mechanisms are not explicitly included. Clearly, factors that affect growth such as nutrients and light for plants should affect the parameters, but *a priori*, it is not obvious how. The most common approach is to consider enrichment of the environment to be reflected in an increase in the carrying capacity K. Note that the model formulation implies that this would not affect the maximum growth rate r at low population densities. Another formulation for logistic growth is:

$$\frac{dA_i}{dt} = rA - cA^2.$$

This can be rewritten as the classic equation given earlier, substituting K for r/c. Thus, it is the same formulation but with subtly different interpretation. Most importantly, r now has a different meaning, as increasing the value of r in the second formulation does not lead to an increase in the equilibrium density (K) in the first equation, whereas in the second equation, the equilibrium population density increases linearly with r. In the alternative equation, c is a competition coefficient. Note that in both equations, gains and losses (birth and death; growth and respiration) are lumped into r.

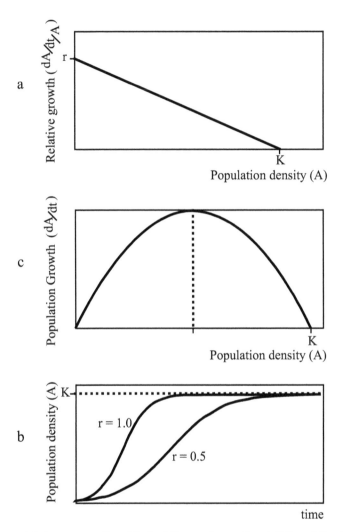

FIGURE A.1. Some properties of the logistic growth equation: (a) The relative (per capita) growth rate ($dA/dt/A$) decreases linearly with the population density. (b) The growth rate (dA/dt) of the total population has a maximum at half the carrying capacity. (c) Population density of a growing population evolves in a sigmoidal way over time.

Parameters can be chosen freely to experiment. Thinking of algal growth in a lake, reasonable choices are $r = 1$ (d^{-1}) and $K = 10$ ($mg\,l^{-1}$) (equivalent to $c = 0.1$ in the second equation).

A.2 Allee Effect

Some populations may have a threshold density below which they go into free fall toward extinction. In the main text (section 2.1), different mechanisms that may cause this are explained. One of the many possible equations that can describe such an Allee effect in a single population is[1]:

$$\frac{dA}{dt} = r\,A\left(1 - \frac{A}{K}\right)\left(\frac{A - a}{K}\right).$$

The only difference from the classic logistic growth equation is the multiplication by the last term $[(A - a)/K]$, which implies that the population growth becomes zero not only at carrying capacity ($A = K$), but also at the Allee extinction threshold a ($A = a$). Below this critical density ($A < a$), net growth is negative and the population goes extinct.

Parameters can be chosen freely, but obviously a should be smaller than K to obtain the typical Allee effect.

A.3 Overexploitation

Consumption (or more generally, exploitation) can also lead to alternative attractors in a plant population (or more generally, exploited population; section 2.2). As more food is available, animals may respond numerically (by reproducing more) but also functionally (by eating more; figure A.2). Holling[2] proposed the now classic formulations for the functional response that he tested elegantly using a blindfolded assistant collecting pieces of sandpaper attached to a table. The simplest realistic version that is commonly used is the *type II functional response*, assuming that the per capita consumption rate (*cons*)

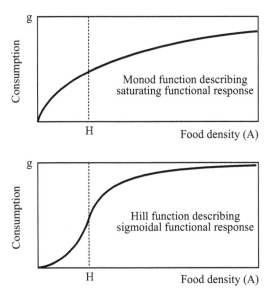

FIGURE A.2. Increase of consumption with food density can follow a simple saturating pattern (top panel) or be sigmoidal (bottom panel).

asymptotically approaches a maximum (g) as food density (A) increases. This can be written as a *Monod equation* with half-saturation constant of H:

$$cons = g \, \frac{A}{A + H}.$$

Another possibility is that consumption rises more steeply around some threshold value. Such a *type III functional response* may arise if the animal switches to alternative food when the density of food source F becomes too low or if a certain part of F cannot be reached (for example, the roots of plants or part of the food population hiding in crevices). A convenient way to formulate that is through a *Hill equation*:

$$cons = g \, \frac{A^p}{A^p + H^p}.$$

Here H is still a half-saturation coefficient, but the power p makes the curve sigmoidal. The higher p, the steeper the increase. At very high values of p (say, $p : 10$), the curve becomes a step-function jumping from zero to g around H. Note that the Monod function is a special case of the Hill function that is obtained by setting p to 1.

To compute the total consumption of a consumer population, one has to multiply either of the preceding equations by the amount of consumers (c).

Figures 2.11 and 2.12 consist of curves for production combined with curves for consumption. These can be approximated by plotting the logistic growth equation and a type III functional response. Playing with the parameters, one can explore under which conditions alternative equilibria exist. The catastrophe fold in the third figure consists of the solutions ($\mathrm{d}A/\mathrm{d}t = 0$) of the differential equation describing the dynamics of the food population as a function of growth and consumption losses:

$$\frac{\mathrm{d}A}{\mathrm{d}t} = r A \left(1 - \frac{A}{K}\right) - g\,c\,\frac{A^p}{A^p + H^p}\,.$$

Simulations with this equation may illustrate the hysteresis discussed in the main text and show that depending on initial conditions, the model may stabilize in an underexploited state or an overexploited state for parameter values that allow alternative equilibria.

A.4 Competition between Two Species

Alternative stable states may also arise simply from competition between two species. The general rule here is that competition can lead to alternative stable states if it is better to have individuals of the same species around than individuals of the other species. (Intraspecific competition is less severe than interspecific competition.) This implies a positive feedback in developing toward a monoculture of either species (the alternative attractors). The implications can be understood well using a classic graphical method to analyze interaction between two species. It is a tricky approach to understand at first, but

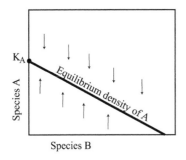
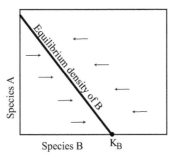

FIGURE A.3. Competition of two species can be seen as a process in which they affect the carrying capacity of each other. Increasing abundance of species B reduces the carrying capacity of A, and hence the equilibrium density to which species A will settle (left panel). Similarly, the equilibrium density to which species B will settle depends on the density of species A (right panel).

once you see how it works, it becomes a powerful way of analyzing the interaction of all kinds of variables, ranging from interaction between predators and their prey to the feedback between the Earth's temperature and greenhouse gas concentrations.[3] The idea is to draw the line at which one species is in equilibrium (the zero-growth *isoclines*) as a function of the other species and vice versa (figure A.3). Such isoclines can be produced, for instance, from the simplest model for competition between two species (a *Lotka-Volterra model*):

$$\frac{dA}{dt} = r_a A - c_a A^2 - c_{ab} A B,$$

$$\frac{dB}{dt} = r_b B - c_b B^2 - c_{ba} B A.$$

Note that the first part of the growth equation of species A is simply the logistic equation. However, in addition to the intraspecific competition ($c_a A$), species A now suffers from competition by the other species ($c_{ab} B$). The equation for the growth of species B is analogous to that for species A.

The zero-growth isoclines plotted in figure A.3 and figure A.4 are solutions of these growth equations ($dA/dt = 0$ and $dB/dt = 0$). It can be seen from the isoclines that the model assumes that the equilibrium

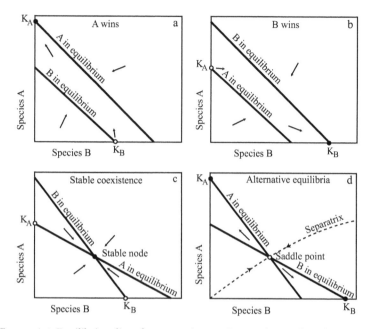

FIGURE A.4. Equilibrium lines for competing species A and B combined reveal the equilibria resulting from the competition. Dots are stable equilibria, whereas open circles represent unstable equilibria.

density of each species declines linearly with increasing density of its competitor (figure A.3). The steepness of the decline depends of course on the intensity of competition. If the competitor has a somewhat different niche, it may be that its suppressing effect is minor.

To see the effect of the competition, one can plot the equilibrium lines of the two species together (figure A.4). If one equilibrium line is entirely above the other, the species with the highest equilibrium wins. If the two lines intersect, this is a point at which both species are at equilibrium. However, this point of coexistence may be either a stable or an unstable equilibrium. It can be seen from the two bottom panels in figure A.4 why stable coexistence requires that interspecific competition is relatively weak, whereas alternative equilibria arise if interspecific competition is stronger than intraspecific competition (implying that the decline in equilibrium density as a function of the density of the competitor is steep).

To see the difference in response to environmental conditions, imagine what happens if the environment changes in a way that is good for species A but has no direct effect on species B. This implies that the equilibrium line for species A in the diagram moves up (as K_A increases), whereas the line for B remains unaltered. If we start in a situation in which species B would win (figure A.4b), this will eventually turn the system into a situation in which A will win (figure A.4a). However, the transition depends on whether coexistence is stable or unstable (figure A.5).

In the case of stable coexistence (figure A.4c; figure A.5, upper panel), as soon as the lines cross, the resulting stable point at which both species are in equilibrium will move gradually to the upper left, implying that species B becomes gradually less abundant in the co-existence equilibrium. When the intersection point eventually reaches the vertical axis, species B goes extinct, and a monoculture of A remains. This is a so-called *transcritical bifurcation*. It is not a catastrophic bifurcation, as the change for species B from rare to extinct is only a small (albeit important) one.

Now consider what happens if the same environmental change occurs in a system with alternative equilibria (figure A.4d; figure A.5, lower panel). Suppose that the system is in the equilibrium dominated by species B (at K_B). If the lines cross, a further increase in K_A will cause the unstable intersection point to slide to the lower right. However, the B monoculture equilibrium remains stable as long as the unstable intersection point does not hit it. Nonetheless, it can be seen that the resilience of this monoculture decreases. The unstable saddle point is on the border between the basins of attraction of the two alternative monoculture equilibria. As the saddle moves toward the B monoculture point, the attraction basin of this equilibrium thus shrinks. This implies that moderate invasions of species A may be enough to trigger a shift to the A monoculture state. Eventually, if the equilibrium line of A moves enough to let the saddle collide with the B monoculture equilibrium (K_B), this monoculture becomes unstable. Even the slightest invasion by A will now cause a runaway shift to a monoculture of A at K_A.

Note that as long as A remains absent, the shift away from the B monoculture will not occur. Such an unstable point in which one

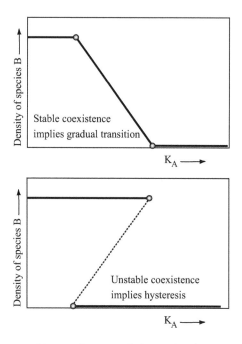

Figure A.5. In competition, environmental change that favors species A but has no direct effect on species B can lead to a gradual (left) or catastrophic (right) transition, depending on the relative strength of interspecific versus intraspecific competition (see text for explanation).

variable is zero is sometimes called a *trivial equilibrium*. This is probably a common situation in nature, as illustrated by the fact that biological invasions are frequently successful. Many species are not absent because they cannot coexist with the rest of the community, but rather because they never arrived.

A.5 Multispecies Competition

Critical transitions in hypothetical communities with many species can be analyzed in an expansion of the competition model discussed in the preceding section. The formulation becomes[4]:

$$\frac{dN_i}{dt} = r\,N_i \left(1 - \frac{\sum_j \alpha_{i,j}\,N_j}{K_i^*} \right) + u, \qquad i = 1, 2, \ldots, n;\ (\alpha_{i,i} = 1,$$

$$K_i^* = K_i\,(1 + M\,\eta_i).$$

This formulation allows one to explore the effect of external forcing, as the carrying capacity of each species is made subject to a hypothetical environmental factor (M) through a randomly assigned sensitivity coefficient (η_i). A small immigration factor (u) is used to prevent unrealistically low species biomasses.

A.6 Predator – Prey Cycles

As a basis for the analyses of limit cycles (section 3.1), we use a simple two-equation predator-prey model that is known to be a reasonable description of what happens in lake plankton[5]:

$$\frac{dA}{dt} = r\,A\left(1 - \frac{A}{K}\right) - g_z\,Z\frac{A}{A + h_a}\,.$$

$$\frac{dZ}{dt} = e_z g_z\,Z\left(1 - \frac{A}{A + h_a}\right) - m_z\,Z\,.$$

Parameter values and units for this model and the following expansions on it are listed in table 1. The basic growth of algae (A) is logistic with a maximum growth rate (r) and a carrying capacity (K). An extra term is added to account for the consumption by zooplankton. This consumption depends on the amount of zooplankton (Z) and its maximum consumption rate (g_z) and on the phytoplankton density. The latter dependence is formulated as a Monod function representing a simple saturating functional response with a fixed half-saturation value (h_a). The zooplankton population converts the ingested food to growth with a certain efficiency (e_z) and suffers losses due to respiration and mortality at a fixed rate (m_z). For the default parameter setting, this model has a limit cycle.

TABLE 1. Overview of Symbols Used in the Equations and Their Dimensions, Default Values, and Meanings

Symbol	Units	Value	Definition
ε	—	0.7	Intensity of seasonal variation in light and temperature
A	mg l^{-1}	—	Concentration of phytoplankton
d	day^{-1}	—	Fraction of volume exchanged between parts with and without Z
e_z	g g^{-1}	—	Efficiency of conversion of food into growth of zooplankton
G_f	mg l^{-1} day^{-1}	—	Maximum zooplankton consumption of entire fish community
g_z	g g^{-1} day^{-1}	0.4	Maximum grazing rate of zooplankton
h_a	mg l^{-1}	0.6	Half-saturation concentration of algae for Z functional response
h_z	mg l^{-1}	1	Half-saturation concentration of Z for fish functional response
i	g g^{-1} day^{-1}	0.01	Inflow of phytoplankton from ungrazed parts
K	mg l^{-1}	10	Carrying capacity for phytoplankton
l	day^{-1}	0.1	Loss rate of phytoplankton
m_z	day^{-1}	0.15	Mortality rate of zooplankton
q	—	—	Fraction of total lake volume occupied by zooplankton concentrations
r	day^{-1}	0.5	Maximum growth rate of phytoplankton
Z	mg l^{-1}	—	Concentration of large herbivorous zoo plankton

A.7 The Hopf Bifurcation

To see the model shift from stable to oscillating (figure 3.3), we mimic the effect of enrichment manipulating K. Increasing the carrying capacity, starting from a very low level, algal density at equilibrium equals K until the level beyond which zooplankton growth becomes positive is reached. As a result, algal density stays constant with further enrichment while zooplankton density increases. When the system is enriched further, the system starts oscillating along a limit cycle at a next critical point. More precisely, as explained in the main text (section 3.1 and figure 3.3), what happens at this Hopf bifurcation is that the stable equilibrium becomes unstable and surrounded by a stable limit cycle.

A.8 Stabilization by Spatial Heterogeneity

As an illustration of how spatial structure can affect the potential of a system to oscillate, we expand the simple model of the interaction between zooplankton and algae. Obviously, modeling the spatial processes in detail is rather complex. A way to simplify while preserving the essential feature of spatial aggregation is to consider zooplankton

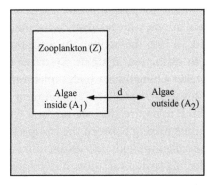

FIGURE A.6. Simple spatial structure used to probe the effect of spatial heterogeneity in a model of plankton cycles. Zooplankton (Z) is confined to one part of the space. Algae grow both inside (A_1) and outside (A_2) the zooplankton compartment and diffuse (d) between both parts of space. (From reference 6.)

to be aggregated in one part of the lake, while algae are present homogeneously throughout the lake. Thus, we have two imaginary compartments (figure A.6). In the first compartment, zooplankton grazes down the local subpopulation of algae (A_1), while in the second compartment, the algae (A_2) are predation-free. Between the compartments, we define an exchange of a fraction d of the lake volume per day. This can be thought of as water with algae moving through the areas where zooplankton is concentrated, but it could equally represent the effect of the movement of the aggregations of zooplankton through the lake.

The model formulations to mimic this situation are:

$$\frac{dA_1}{dt} = r A_1 \left(1 - \frac{A_1}{K}\right) - g_z Z \frac{A_1}{A + h_a} + \frac{d}{f}(A_2 - A_1),$$

$$\frac{dA_2}{dt} = r A_2 \left(1 - \frac{A_2}{K}\right) - \frac{d}{1-q}(A_2 - A_1),$$

$$\frac{dZ}{dt} = e_z g_z Z \frac{A_2}{A_1 + h_a} - m_z Z.$$

The effect of the mixing rate and the fraction of the lake occupied by zooplankton on the dynamics of the model can summarized in a few graphs (figure A.7). If there is no mixing (I), the populations in the compartment in which zooplankton and algae occur together go through strong oscillations, whereas algae in the ungrazed part of the lake are unaffected. In fact, the oscillations drive the algae and their consumers close to extinction at times. Recovery from such near-extinction events takes a long time, which explains the long period of the oscillations. This type of oscillation is common in models, but if oscillations are found in nature, they are rarely as wild as this. Simulations in which some mixing between the compartments is allowed illustrate that the oscillations with near-extinction events may well be an artifact of neglecting spatial variation. A slight mixing (II) is sufficient to change the dynamics in the grazed part completely. The system still oscillates, but the amplitude of the cycles is reduced, and the period of the oscillations becomes closer to what is observed in the field. The algae in the ungrazed part now show a mild oscillation too,

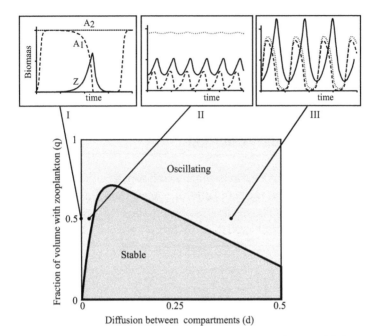

FIGURE A.7. Bifurcation graph of the spatial zooplankton–algae model showing the effect of the fraction of the volume occupied by zooplankton (q) and the diffusion rate of algae between the two parts of space (d). The curve represents the Hopf bifurcations and defines the borderline in parameter space between oscillatory and stationary behavior. A stable equilibrium state exists when the grazed fraction of the space (q) is sufficiently small and the exchange rate (d) is intermediate. Stabilization becomes increasingly difficult if zooplankton occupies a larger part of space. Time plots illustrate the behavior of the model at different levels of mixing. (Adapted from reference 6.)

driven by an exchange with the grazed part. A further increase of mixing causes the limit cycle to shrink and finally collide with the unstable equilibrium point to produce a stable equilibrium point. This is a Hopf bifurcation (now reversed).

So far, mixing over the spatially structured system thus appears stabilizing. The surprise comes if we increase the mixing rate further. The system then passes through another Hopf bifurcation and becomes oscillatory again. Although this may seem counterintuitive at first glance, it makes sense if one imagines the extreme situation of very strong mixing. We then approach the limit case of a completely mixed volume again. The dynamics of both algal subpopulations become increasingly alike and the density of zooplankton in its refuge becomes very high because of the inflow of food from the rest of the lake.

A.9 Basin Boundary Collision

To see an example of a basin boundary collision, we add a top-predator (in this case, fish) to the zooplankton − algae model discussed in the preceding section. The algal equation remains the same, and to account for the effect of predation by fish on zooplankton, we simply add an extra loss term to their growth equation:

$$\frac{dA}{dt} = r\,A\left(1 - \frac{A}{K}\right) - g_z Z\,\frac{A}{A + h_a} + i\,(K - A),$$

$$\frac{dZ}{dt} = e_z g_z Z\,\frac{A}{A + h_a} - m_z Z - F\,\frac{Z^2}{Z^2 + h_z^2}.$$

Many behavioral scenarios can arise from this model. To see an example of a basin boundary collision, one can start at a fish density (G_f) of zero and then gradually increase this parameter. If the plankton interaction has a cyclic attractor and the amplitude of the limit cycle is large enough to hit the border of the attraction basin (figure 3.8), the system collapses into the green state. This is the basin boundary collision (or *homoclinic bifurcation*).

A.10 Periodic Forcing

As an example of the potential effects of periodic forcing, we take a model of the effect of seasons on lake plankton.[7] We assume light, temperature, and fish predation to vary over the seasons. Light affects the carrying capacity (K) of algae, and temperature affects the parameters related to the metabolism of algae, zooplankton, and fish (r, g, m, G_f). For simplicity, we neglect many subtleties such as the exact shapes of temperature and light dependence of the organisms and simply mimic the effect of seasons by multiplying each of those parameters by a seasonal impact, σ_t, which is a periodic function of time t (days):

$$\sigma_{(t)} = \frac{1 - \varepsilon \cos\left(\dfrac{2\pi t}{365}\right)}{1 + \varepsilon}.$$

where $t = 0$ stands for the first of January. In this formulation, the minimum value of each parameter (that is, the value in the middle of the winter) is equal to its maximum (in summer) multiplied by $(1 - \varepsilon)/(1 + \varepsilon)$. Thus, the summer maximum of a parameter corresponds to the default value, and determines the amplitude of seasonal change.

In addition to the temperature-induced variation, fish predation pressure (G_f) should show a seasonal variation due to the reproductive cycle. Assuming this cycle to be sinusoidal and in phase with the variation in temperature and light, we can include the effect simply by multiplying G_f with an extra seasonal impact $(\sigma_{(t)})$. Thus, the complete seasonal model becomes:

$$\frac{dA}{dt} = \sigma_t \, r A \left(1 - \frac{A}{\sigma_{(t)}K}\right) - Z \sigma_{(t)} g_z \frac{A}{A + h_A} + d(\sigma_{(t)}K - A),$$

$$\frac{dZ}{dt} = e_z g_z Z \frac{A}{A + h_a} - m_z Z - F \frac{Z^2}{Z^2 + h_z^2}.$$

The analysis and interpretation of different types of attractors and bifurcations of this seasonal model are quite complex,[7] but an easy way to investigate its modes of behavior is to simply simulate the dynamics for different parameter settings (figure 7.9).

A.11 Self-Organized Patterns

Models that produce self-organized patterns have been proposed by various authors.[8,9] The essence is always that there is a local positive feedback resulting from a positive effect of plants on local water

TABLE 2. Parameters, Their Default Values, and Units (following reference 9)

Parameter	Description	Default	Unit
c	Conversion of water uptake by plants to plant growth	10	$\text{g mm}^{-1} \text{m}^{-2}$
g_{max}	Maximum specific water uptake	0.05	$\text{mm g}^{-1} \text{m}^2 \text{d}^{-1}$
k_1	Half-saturation constant of specific plant growth and water uptake	5	mm
d	Specific loss of plant density due to mortality	0.25	d^{-1}
D_p	Plant dispersal	0.1	$\text{m}^2 \text{d}^{-1}$
α	Maximum infiltration rate	0.2	d^{-1}
k_2	Saturation constant of water infiltration	5.0	g m^{-2}
W_0	Water infiltration rate in absence of plants	0.2	—

(and/or nutrient) availability, while at farther distances, there is a negative feedback resulting from competition between the plant patches for water (and/or nutrients). As an example we take the model by Rietkerk et al.[9] consisting of the three partial differential equations. The model allows for lateral displacement of water and vegetation through diffusion. Plant biomass V_b is described by the following partial differential equation. (Parameter descriptions can be found in table 2.)

$$\frac{\partial V_b}{\partial t} = c\, g_{max}\left(\frac{W}{W + k_1}\right)V_b - dV_b + D_{V_b}\,\Delta V_b.$$

The first term corresponds to plant growth, calculated as a function of water availability W, plant biomass V_b, and maximum water uptake by plants g_{max}. The second term of the equation represents plant mortality d, and the third term represents plant dispersion, which is modeled as a diffusion term with diffusion constant D_{V_b}. Soil water W is modeled as:

$$\frac{\partial W}{\partial t} = \alpha O \frac{V_b + k_2 W_0}{V_b + k_2} - g_{max}\left(\frac{W}{W + k_1}\right) V_b - r_w W + D_w \Delta W.$$

This equation describes the infiltration of water, the water loss due to uptake and evaporation and drainage, and the diffusion of soil water with its respective terms.

Surface water O is described as:

$$\frac{\partial O}{\partial t} = P - \alpha O\left(\frac{V_b + k_2 W_0}{V_b + k_2}\right) + D_O \Delta O.$$

The terms in this equation represent rainfall, the loss of water due to infiltration, and the diffusion of surface water, respectively. Simulations with this model on a grid of coupled cells produces the different patterns shown.

A.12 Alternative Stable States in Shallow Lakes

The essence of the mechanism causing alternative stable states in shallow lakes is that vegetation reduces turbidity, whereas turbidity hampers vegetation growth. The graphical model presented in the section 7.1 corresponds to a simple set of equations.[10] An inverse Monod function is used to describe the vegetation effect on the equilibrium turbidity (E_{eq}):

$$E_{eq} = E_0 \frac{h_v}{h_v + V},$$

in which V is the area of the lake covered with vegetation, and E_0 is the turbidity in the absence of vegetation. The effect of extinction on vegetation is modeled as a Hill function:

$$V_{eq} = \frac{h_E{}^p}{h_E{}^p + E^p},$$

If we assume that turbidity and vegetation biomass dynamically approach their equilibrium values simply in a logistic way, the preceding equations describing the equilibria can be transformed into the following differential equations:

$$\frac{dE}{dt} = r_E E \left(1 - \frac{E}{E_{eq}}\right),$$

$$\frac{dV}{dt} = r_V V \left(1 - \frac{V}{V_{eq}}\right).$$

A.13 Floating Plants

The asymmetry in competition between free-floating and rooted submerged plants (section 7.3) has three essential features: (1) floating plants have primacy for light, whereas submerged plants (2) can grow at lower water column nutrient concentrations and (3) can reduce water column nutrients to lower levels. A simple model to explore the potential implications of this specific asymmetry is shown here[10]:

$$\frac{da}{dt} = a \left(r_a \frac{1}{1 + q_a (t_a a + t_c c)} \frac{P - c - a}{h + P - c - a} - d_a - f\right),$$

$$\frac{dS}{dt} = r_s S \left(\frac{n}{n + h_s}\right) \left(\frac{1}{1 + a_s S + bF + W}\right) - l_s S.$$

Change over time of the biomass of floating plants, F, and submerged plants, S, is modeled as a function of their maximum growth rates, r_f and r_s, modified by nutrient and light limitation, and of their losses, l_f and l_s, due to processes such as respiration and various mortality factors. Nutrient limitation is a saturating function of the total inorganic nitrogen concentration, n, in the water column, which is assumed to be a decreasing function of plant biomass:

$$n = \frac{N}{1 + q_s S + q_f F},$$

where the maximum concentration N in the absence of plants depends on the nutrient loading of the system, and the parameters q_s and

Table 3. Default Values and Dimensions of
Parameters and Variables of the Floating
Plant Model

Parameter/Variable	Value	Units
F	—	g dw m^{-2}
S	—	g dw m^{-2}
N	—	mg N l^{-1}
n	—	mg N l^{-1}
a_f	0.01	$(\text{g dw m}^{-2})^{-1}$
a_s	0.01	$(\text{g dw m}^{-2})^{-1}$
b	0.02	$(\text{g dw m}^{-2})^{-1}$
h_f	0.2	mg N l^{-1}
h_s	0.0	mg N l^{-1}
l_f	0.05	day^{-1}
l_s	0.05	day^{-1}
q_f	0.005	$(\text{g dw m}^{-2})^{-1}$
q_s	0.075	$(\text{g dw m}^{-2})^{-1}$
r_f	0.5	day^{-1}
W	0	—
r_s	0.5	day^{-1}

q_f represent the effect of submerged and floating plants on the nitrogen concentration in the water column. Light limitation is formulated in a simple fashion, where $1/a_f$ and $1/a_s$ are the densities of floating and submerged plants at which their growth rates becomes reduced by 50% due to intraspecific competition for light. In addition to this intraspecific competition, irradiation of submerged plants is reduced by light attenuation in the water column W and by shading by floating plants scaled by the parameter b. Default values of parameters to produce the figures shown in the main text are given in table 3.

A. 14 Contingency in Behavior

To model the effect of contingency in behavior in a simple way, we suppose that for each individual, there are simply two modes of *opinion* or *attitude* with respect to a problem: active $(+1)$ or passive (-1). It takes effort to be active, but activation also generates pressure to authorities in the direction of one's own interest as well as a "warm glow" feeling[12] that one is doing "the right thing." Let $\tilde{U}_{(+)}$ denote the perceived payoff or utility to being active and $\tilde{U}_{(-)}$ the utility of being passive. These utilities have a random component to reflect idiosyncrasies across people: $\tilde{U}(a) = U(a) + s\varepsilon(a)$ for action $a = +1, -1$, where $U(a)$ is deterministic, $\varepsilon(a)$ is a random variable, and s scales the variance. It turns out that if $\varepsilon(a)$ is independently and identically distributed across people and actions, we can apply the law of large numbers and compute the probability P of action a as a function of $U(a)$, a, and s:

$$P(a) = \frac{e^{\frac{U(a)}{s}}}{e^{\frac{U(+1)}{s}} + e^{\frac{U(-1)}{s}}}.$$

We now introduce peer group "social pressure" effects. We define $n_t(a)$ as being the probability P of action a at time t, and the overall tendency for action as:

$$A_t = n_t(+1) - n_t(-1).$$

and assume the perceived utility for person i at time t of taking a certain action to be affected also by the cost $c(a_{i,t} - A_t)^2$ of deviating from the overall group tendency, obtaining:

$$V_t(a_i) = U_t(a_{i,t}) - c(a_{i,t} - A_t)^2.$$

Then adapting the probability function and replacing U with V, we have:

$$A_t = T\left(\frac{h_t + 2cA_{t-1}}{s}\right)$$

$$h_t = \left(\frac{U_{t(+1)} - U_{t(-1)}}{2} \right)$$

$$T_x = \frac{e^x - e^{-x}}{e^x - e^{-x}},$$

The rationale behind the model is discussed extensively in two papers dealing with paradigm shifts.[13] The figures of the response of public attitude to an increasing problem are obtained by plotting the equilibrium action level $\bar{A}_t$ (for $A_t = A_{t-1}$) as a function of h_t.

Glossary

An explanation of technical terms from dynamical systems theory is given in the following table in heuristic rather than mathematically rigorous terms. More details about these terms can be found in the text.

Adaptive cycle	Heuristic model of a cycle of collapse and reorganization observed in many systems adjusting to changing conditions.
Alternative stable states	Contrasting states to which a system may converge under the same external conditions.
Attractor	A state or dynamic regime to which a model converges given sufficient simulation time.
Basin boundary	Divisions between sets of initial conditions that lead to different attractors.
Basin boundary collision	Event at which internally driven oscillations in the state of the system tip the system into an alternative basin of attraction.
Basin of attraction	Set of initial conditions that lead to a particular attractor.
Bifurcation	Threshold in parameters (or conditions) at which the qualitative behavior of a model changes.
Biomanipulation	Managing the abundance of key species in an attempt to change the overall state of an ecosystem. Used in particular for removal of fish from a lake as a way to promote water clarity.

Catastrophe fold	Folded response-curve of the state of a system to a conditioning factor, implying that for given conditions, the system has alternative stable states.
Catastrophic bifurcation	Bifurcation where an attractor disappears so that the system is forced to move to an alternative attractor.
Catastrophic shift	Substantial shift in the state of a model induced by a tiny perturbation bringing it across a catastrophic bifurcation.
Chaos	More precisely referred to as *deterministic chaos*. Unpredictable fluctuations in models resulting from deterministic rules.
Community	Group of species living together in an ecosystem.
Competitive exclusion	Exclusion of a species by a stronger competitor.
Contingency	Used here to refer to the tendency of individuals to copy the attitude of peers.
Critical slowing down	The tendency that the recovery from disturbances slows down close to a bifurcation.
Equilibrium	Situation in which processes that affect the state of a system precisely balance out so that the system does not change. An equilibrium is *stable* if the system returns to it after a small perturbation and is *unstable* if the system moves away from the equilibrium point after such a perturbation.
Eutrophication	Changes in an ecosystem due to excessive enrichment with nutrients.
External forcing	Influence of factors outside the system on its state (for example, effect of solar radiation on the Earth).
Feedback loop	Set of cause–effect relationships that form a closed loop.
Fold bifurcation	Critical threshold where a stable point is touched by an unstable point that marks the border of the basin of at-

traction. This bifurcation thus marks the abrupt disappearance of an equilibrium.

Functional group — A group of species that perform largely the same function in the ecosystem, such as insect-eaters or nitrogen-fixing plants.

Ghost — The remains of an attractor that has become unstable but still slows down the behavior of the system in its vicinity.

Hysteresis — The tendency of a system to stay in the same state if conditions change. Specifically, in the context of alternative stable states, the term refers to the fact that as conditions are changing, the system remains on the same attractor until a catastrophic bifurcation is reached where the system jumps to the alternative attractor. If the conditions are then changed in the opposite direction, the system stays on the new attractor until another catastrophic bifurcation makes it jump back to the original state. This completes a *hysteresis loop*. The range of conditions between the two bifurcation points is also referred to as the *size* or *width* of the hysteresis.

Isotopes and isotopic signature — Atoms of common elements such as oxygen or nitrogen exist in various forms that are similar for all practical purposes, but differ slightly in weight. Many processes such as evaporation of water or nitrogen fixing by bacteria have a slight preference for the lighter or the heavier forms. Therefore, these processes result in an altered ratio of isotopes known as an *isotopic signature* that can be used to deduce what processes have happened from looking, for instance, at ancient ice or ocean sediments. Also, isotope ratios may help to reconstruct the diet of animals from their body tissue.

Limit cycle — Stable cyclic dynamics of a system generated by internal processes. Such a cycle is an attractor, as starting from different states, the system is pulled toward this limit behavior.

Mechanistic model — Model that explicitly accounts for the mechanisms that are thought to play a role. In contrast, empirical models (such as regression models) describe observed relationships between variables.

Minimal (strategic) model	Model that focuses on a minimal set of mechanisms needed to produce a certain behavior.
Multidimensional system	A system with many variables. To plot the state of the system, one axis (dimension) is needed to represent the value of each of the variables.
Multiple stable states	A situation in which a system has more than one stable state given the same external conditions.
Niche	The way in which a species "makes a living" in an ecosystem. For instance, the niche of a particular bird may be to forage on small flying insects, and a particular plant may be specialized in growing in acidic, wet, nutrient-poor soils.
Nonlinear system	A system in which the dynamics are not linearly dependent on the state. This can lead to phenomena such as thresholds, multiple attractors, cycles, and chaotic dynamics.
Overexploitation	Exploitation (or harvest or consumption) at a rate that compromises the overall productivity of the exploited population.
Periodic forcing	Influence of changes in conditions that oscillate with a fixed period, such as seasonal or diurnal variations in temperature and light.
Positive feedback	A chain of effects through which something has a positive effect on itself. For instance, climate warming in some regions may cause snowmelt. The dark vegetation thus uncovered absorbs more solar radiation than the snow, and this leads to further warming.
Quasi-periodic	An oscillating system in which the behavior is not precisely periodic due to changing amplitude and period. In contrast with chaotic systems, quasi-periodic systems do not have the sensitivity to initial conditions that causes long-term unpredictability.

Regime shift
: A relatively sharp change from one regime to a contrasting one, where a regime is a dynamic "state" of a system with its characteristic stochastic fluctuations and/or cycles.

Repellor
: The opposite of an attractor. Thus a state or dynamic regime from which a model moves away. Repellors mark the border between alternative basins of attraction.

Resilience
: The magnitude of disturbance that a system can tolerate before it shifts into a different state.

A more liberal definition used for social–ecological systems is this: The capacity of a system to absorb disturbance and reorganize while undergoing change so as to still retain essentially the same function, structure, identity, and feedbacks.

Resistance
: Force needed to cause a certain change in a system.

Runaway change
: Change that is accelerating and self-propelling due to a positive feedback.

Saddle
: A particular unstable equilibrium that attracts in some directions but repels in others.

Self-organized patterns
: Patterns that emerge automatically from the interaction between many units.

Shallow lake
: A lake of only a few meters deep where the water column is usually well mixed. Most lakes are of this type.

Slow–fast cycle
: A limit cycle with contrasting slow and fast phases that can be interpreted as internally generated periodic shifts between alternative stable states.

Spatial heterogeneity
: Variation of factors in space so that sites differ.

Stable equilibrium
: Attractor point.

Strange attractor
: Attractor on which the system shows chaotic behavior.

Threshold
: A point where the system is very sensitive to changing conditions.

Transient	Behavior of a dynamical system on its way to an attractor.
Unstable equilibrium	Repellor point.
Utility	Here, the benefits that humans can get from an ecosystem.

Notes

Chapter 1

1. M. Nystrom, C. Folke, and F. Moberg, *Trends Ecol. Evol.* **15** (10), 413 (2000); T. P. Hughes, *Science* **265**, 1547 (1994); N. Knowlton, *Am. Zool.* **32**(6), 674 (1992); T. J. Done, *Hydrobiologia* **247**(1–2), 121 (1991); L. J. McCook, *Coral Reef* **18**, 357 (1999); D. R. Bellwood, T. P. Hughes, C. Folke et al., *Nature* **429**(6994), 827 (2004).

2. P. deMenocal, J. Ortiz, T. Guilderson et al., *Quaternary Science Reviews* **19**(1–5), 347 (2000).

3. M. Gladwell, *The Tipping Point* (Little, Brown, and Co., New York, 2000).

4. R. Adler, *Nature* **414**(6863), 480 (2001).

5. J. Diamond, *Collapse: How Societies Choose to Fail or Survive* (Viking Adult, New York, 2004).

6. G. Schwartz and J. J. Nicols, *After Collapse, The Regeneration of Complex Societies* (The University of Arizona Press, Tucson, 2006).

7. T. Heyerdahl, *In the Footsteps of Adam: An Autobiography*, p. 320 (Little, Brown, and Co., London, 1998).

8. J. A. Tainter, *The Collapse of Complex Societies: New Studies in Archaeology* (Cambridge University Press, Cambridge, UK, 1988); M. A. Janssen, T. A. Kohler, and M. Scheffer, *Current Anthropology* **44**(5), 722 (2003).

9. S. R. Carpenter, *Regime Shifts in Lake Ecosystems: Pattern and Variation* (Ecology Institute, Oldendorf/Luhe, Germany, 2003).

10. S. A. Forbes *Bulletin of the Scientific Association (Peoria, IL)*, 77 (1887).

Part I

Chapter 2

1. R. Thom, *Structural Stability and Morphogenesis: An Outline of a General Theory of Models* (Addison-Wesley, Reading, MA, 1993).

2. R. C. Lewontin, "The Meaning of Stability," in *Diversity and Stability in Ecological Systems*, Brookhaven Symposiums in Biology, vol. 22, pp. 13–24 (1969).

3. C. S. Holling, *Annu. Rev. Ecol. Syst.* **4**, 1 (1973).

4. R. M. May, *Nature* **269** (5628), 471 (1977).

5. C. S. Holling, *Annu. Rev. Ecol. Syst.* **4**, 1 (1973).

6. G. J. Van Geest, F. Roozen, H. Coops et al., *Freshwater Biol.* **48**(3), 440 (2003).

7. R. Thom, *Structural Stability and Morphogenesis: An Outline of a General Theory of Models* (Addison-Wesley, Reading, MA, 1993).

8. D. L. DeAngelis, W. M. Post, and C. C. Travis, *Positive Feedback in Natural Systems* (Springer-Verlag, New York, 1986).

9. M. Holmgren, M. Scheffer, and M. A. Huston, *Ecology* **78**(7), 1966 (1997).

10. J. B. Wilson and A.D.Q. Agnew, *Adv. Ecol. Res.* **23**, 263 (1992).

11. M. Holmgren and M. Scheffer. *Ecosystems* **4**(2), 151 (2001).

12. M. Scheffer, R. Portielje, and L. Zambrano, *Limnol. Oceanogr.* **48**, 1920 (2003); J. Van de Koppel, P.M.J. Herman, P. Thoolen et al., *Ecology* **82**(12), 3449 (2001).

13. M. Scheffer, S. H. Hosper, M. L. Meijer et al., *Trends Ecol. Evol.* **8**(8), 275 (1993).

14. I. Noy-Meir, *J. Ecol.* **63** (2), 459 (1975).

15. S. Bowles, S. N. Durlauf, and K. Hoff, *Poverty Traps* (Princeton University Press, Princeton, NJ, 2006).

16. S. Bowles and H. Gintis, *J. Econ. Perspect.* **16** (3), 3 (2002).

17. M. M. Holland, C. M. Bitz, and B. Tremblay, *Geophys. Res. Lett.* **33**, 23 (2006).

18. M. I. Budyko, *Tellus* **21**(5), 611 (1969).

19. C. Punckt, M. Bolscher, H. H. Rotermund et al., *Science* **305**(5687), 1133 (2004).

20. S. D. Mylius, K. Klumpers, A. M. de Roos et al., *Am. Nat.* **158**(3), 259 (2001); S. Diehl and M. Feissel, *Am. Nat.* **155**(2), 200 (2000).

21. C. Walters and J. F. Kitchell, *Can. J. Fish. Aquat. Sci.* **58**(1), 39 (2001); A. M. De Roos, L. Persson, and H. R. Thieme, *Proc. Royal Society of London—Biological Sciences* **270**, 611 (2003).

22. I. Hanski, J. Poyry, T. Pakkala et al., *Nature* **377**, 618 (1995).

23. B. J. Crespi, *Trends Ecol. Evol.* **19**(12), 627 (2004).

24. K. Taylor, *Am. Sci.* **87**(4), 320 (1999).

Chapter 3

1. M. L. Rosenzweig, *Science* **171**, 385 (1971).

2. S. Rinaldi and M. Scheffer, *Ecosystems* **3**(6), 507 (2000).

3. D. Ludwig, D. D. Jones, and C. S. Holling, *J. Anim. Ecol.* **47**(1), 315 (1978); C. S. Holling, *Memoirs of the Entomological Society of Canada* **146**, 21 (1988).

4. M. Scheffer, *Ecology of Shallow Lakes*, 1st ed. (Chapman and Hall, London, 1998).

5. W. B. Cutler, *Am. J. Obstet. Gynecol.* **137**(7), 834 (1980).

6. J. Vandermeer, L. Stone, and B. Blasius, *Chaos Solitons Fractals* **12** (2), 265 (2001).

7. A. R. Yehia, D. Jeandupeux, F. Alonso et al., *Chaos* **9**(4), 916 (1999).

8. J. Gleick, *Chaos: Making a New Science* (Penguin Books, New York, 1988).

9. J. Huisman and F. J. Weissing, *Nature* **402**(6760), 407 (1999).

10. R. M. May, *Nature* **261**, 459 (1976).

11. T. D. Rogers, *Progress in Theoretical Biology* **6**, 91 (1981).

12. S. Smale, *J. Math. Biol.* **3**, 5 (1976).

13. E. H. Van Nes and M. Scheffer, *Am. Nat.* **164**(2), 255 (2004).

14. M. Scheffer, *J. Plankton Res.* **13**, 1291 (1991).

15. M. Scheffer, S. Rinaldi, and Y. A. Kuznetsov, *Can. J. Fish. Aquat. Sci.* **57**(6), 1208 (2000).

16. J. Vandermeer and P. Yodzis, *Ecology* **80**, 1817 (1999).

17. J. Huisman and F. J. Weissing, *Am. Nat.* **157**(5), 488 (2001).

Chapter 4

1. B. D. Malamud, G. Morein, and D. L. Turcotte, *Science* **281** (5384), 1840 (1998).

2. P. Bak, *How Nature Works: The Science of Self-Organized Criticality* (Copernicus Books, New York, 1996).

3. P. Bak, *Phys. Rev. Lett.* **59**, 381 (1987).

4. R. V. Sole, S. C. Manrubia, M. Benton et al., *Nature* **388**(6644), 764 (1997).

5. P. Bak and K. Sneppen, *Phys. Rev. Lett.* **71**(24), 4083 (1993).

6. M.E.J. Newman, *J. Theor. Biol.* **189**(3), 235 (1997).

7. J. van de Koppel, D. van der Wal, J. P. Bakker et al., *Am. Nat.* **165**(1), E1 (2005).

8. J. von Hardenberg, E. Meron, M. Shachak et al., *Phys. Rev. Lett.* **8719**(19), Art. No. 198101 (2001).

9. R. HilleRisLambers, M. Rietkerk, F. Van den Bosch et al., *Ecology* **82**(1), 50 (2001); J. Van de Koppel and M. Rietkerk, *Am. Nat.* **163**(1), 113 (2004); M. Rietkerk, M. C. Boerlijst, F. van Langevelde et al., *Am. Nat.* **160**(4), 524 (2002).

10. M. Rietkerk, S. C. Dekker, P. C. de Ruiter et al., *Science* **305** (5692), 1926 (2004).

11. R. Levins, *Am. Sci.* **54**(4), 421 (1966); M. E. Gilpin and T. J. Case, *Nature* **261**, 40 (1976).

12. J. A. Drake, *Trends Ecol. Evol.* **5**(5), 159 (1990).

13. C. L. Samuels and J. A. Drake, *Trends Ecol. Evol.* **12**, 427 (1997).

14. T. J. Case, *Proc. Natl. Acad. Sci. USA* **87**, 9610 (1990); R. Law and R. D. Morton, *Ecology* **77**, 762 (1996).

15. R. A. Matthews, W. G. Landis, and G. B. Matthews, *Environ. Toxicol. Chem.* **15**, 597 (1996).

16. E. H. Van Nes and M. Scheffer, *Am. Nat.* **164**(2), 255 (2004).

17. A. F. Lotter, *Holocene* **8** (4), 395 (1998).

18. R. M. May, *Math. Biosci.* **12**, 59 (1971).

19. S. L. Pimm, *Nature* **307** (5949), 321 (1984).

20. G. E. Hutchinson, *Am. Nat.* **93**(870), 145 (1959).

21. G. E. Hutchinson, *Am. Nat.* **95**, 137 (1961).

22. D. Tilman, *Ecology* **58**, 338 (1977).

23. U. Sommer, *Limnol. Oceanogr.* **29**, 633 (1984); U. Sommer, *Limnol. Oceanogr.* **30**(2), 335 (1985); M. A. Huston, *Biological Diversity: The Coexistence of Species on Changing Landscape* (Cambridge University Press, Cambridge, UK, 1994).

24. J. Huisman and F. J. Weissing, *Nature* **402**(6760), 407 (1999).

25. A. Bracco, A. Provenzale, and I. Scheuring, *Proc. R. Soc. Lond. Ser. B-Biol. Sci.* **267**(1454), 1795 (2000).

26. R. T. Paine, *Am. Nat.* **100**, 65 (1966); J. M. Chase, P. A. Abrams, J. P. Grover et al., *Ecol. Lett.* **5**(2), 302 (2002).

27. C. E. Mitchell and A. G. Power, *Nature* **421**(6923), 625 (2003); M. E. Torchin, K. D. Lafferty, A. P. Dobson et al., *Nature* **421**(6923), 628 (2003).

28. G. Bell, *Am. Nat.* **155**(5), 606 (2000); S. P. Hubbell, *The Unified Neutral Theory of Biodiversity and Biography* (Princeton University Press, Princeton, NJ, 2001).

29. J. Whitfield, *Nature* **417**(6888), 480 (2002).

30. L. W. Aarssen, *Am. Nat.* **122**(6), 707 (1983).

31. M. Loreau, *Oikos* **104**(3), 606 (2004).

32. D. W. Yu, J. W. Terborgh, and M. D. Potts, *Ecol. Lett.* **1**(3), 193 (1998); R. E. Ricklefs, *Oikos* **100**(1), 185 (2003).

33. M. Scheffer and E. H. Van Nes, *Proc. Natl. Acad. Sci. USA* **103**(16), 6230 (2006).

34. R. H. MacArthur and R. Levins, *Am. Nat.* **101**(921), 377 (1967).

35. T. D. Havlicek and S. R. Carpenter, *Limnol. Oceanogr.* **46**(5), 1021 (2001); E. Siemann and J. H. Brown, *Ecology* **80**(8), 2788 (1999); C. S. Holling, *Ecol. Monographs* **62**, 447 (1992).

36. C. R. Allen, *Proc. Natl. Acad. Sci. USA* **103**(16), 6083 (2006).

37. Y. Liu, D. S. Putler, and C. B. Weinberg, *Mark. Sci.* **23**(1), 120 (2004).

38. H. Hotelling, *The Economic Journal* **39**(153), 41 (1929).

39. C. S. Holling, *Annu. Rev. Ecol. Syst.* **4**, 1 (1973).

40. C. S. Holling, in *The Resilience of Terrestrial Ecosystems: Local Surprise And Global Change*, W. C. Clark and R. E. Munn, Eds., pp. 292–317 (Cambridge University Press, Cambridge, UK, 1986).

41. L. Gunderson and C. S. Holling, *Panarchy: Understanding Transformations in Human and Natural Systems* (Island Press, Washington, DC, 2001).

42. D. Ludwig, D. D. Jones, and C. S. Holling, *J. Anim. Ecol.* **47**(1), 315 (1978).

43. B. H. Walker and D. Salt, *Resilience Thinking—Sustaining Ecosystems and People in a Changing World* (Island Press, Washington, DC, 2006).

44. S. R. Carpenter, *Ecology* **77**(3), 677 (1996).

Chapter 5

1. E. Benincà, J. Huisman, R. Heerkloss et al., *Nature* **451**, 822 (2008); K. Kersting, *Verhandlungen Internationale Vereinigung Theoretische Angewandte Limnologie* **22**, 3040 (1985).

2. O. N. Bjornstad and B. T. Grenfell, *Science* **293**(5530), 638 (2001); S. Ellner and P. Turchin, *Am. Nat.* **145**, 343 (1995).

3. S. R. Carpenter, *Regime Shifts in Lake Ecosystems: Pattern and Variation* (Ecology Institute, Oldendorf/Luhe, Germany, 2002).

4. A. Hastings, *Trends Ecol. Evol.* **19**(1), 39 (2004).

5. J. M. Cushing, B. Dennis, R. A. Desharnais et al., *J. Anim. Ecol.* **67**(2), 298 (1998).

6. G. J. Van Geest, H. Coops, M. Scheffer et al., *Ecosystems* **10**(1), 37 (2007).

7. J. H. Connell and W. P. Sousa, *Am. Nat.* **121**(6), 789 (1983); G. D. Peterson, *Clim. Change.* **44**(3), 291 (2000); S. A. Levin, *Ecosystems* **3**(6), 498 (2000); P. S. Petraitis and R. E. Latham, *Ecology* **80**(2), 429 (1999); M. Nystrom and C. Folke, *Ecosystems* **4**(5), 406 (2001).

8. W. S. Gurney and R. M. Nisbet, *J. Anim. Ecol.* **47**(1), 85 (1978); R. Nisbet, W. S. Gurney, W. W. Murdoch et al., *Bio. J. Linn. Soc.* **37**, 79 (1989).

9. M. P. Hassel and R. M. May, *J. Anim. Ecol.* **43**, 567 (1974).

10. A. M. De Roos, E. McCauley, and W. G. Wilson, *Proc. R. Soc. Edin. Sect. B (Biol. Sci.)* **246**, 117 (1991).

11. M. Scheffer, M. S. Van den Berg, A. W. Breukelaar et al., *Aquat. Bot.* **49**, 193 (1994).

12. E. H. Van Nes and M. Scheffer, *Ecology* **86**(7), 1797–1807 (2005).

13. K. S. McCann, *Nature* **405**(6783), 228 (2000).

14. A. R. Ives and S. R. Carpenter, *Science* **317**(5834), 58 (2007).

15. E. P. Odum, *Fundamentals of Ecology* (W.B. Saunders Co., Philadelphia and London, 1953).

16. S. Yachi and M. Loreau, *Proc. Natl. Acad. Sci. USA* **96**(4), 1463 (1999).

17. M. Loreau, *Oikos* **91**(1), 3 (2000).

18. T. Elmqvist, C. Folke, M. Nystrom et al., *Front. Ecol. Environ.* **1**(9), 488 (2003); B. Walker, A. Kinzig, and J. Langridge, *Ecosystems* **2**(2), 95 (1999).

19. T. P. Hughes, *Science* **265**, 1547 (1994).

20. H. T. Dublin, A. R. Sinclair, and J. McGlade, *J. Anim. Ecol.* **59**(3), 1147 (1990).

21. M. Scheffer, M. Holmgren, V. Brovkin et al., *Global Change Biology* **11**(7), 1003 (2005).

22. M. Maslin, *Science* **306**(5705), 2197 (2004).

23. R. D. Vinebrooke, K. L. Cottingham, J. Norberg et al., *Oikos* **104**(3), 451 (2004).

24. A. R. Ives and B. J. Cardinale, *Nature* **429**(6988), 174 (2004).

25. E. H. Van Nes and M. Scheffer, *Am. Nat.* **164**(2), 255 (2004).

26. R. S. Steneck, J. Vavrinec, and A. V. Leland, *Ecosystems* **7**(4), 323 (2004).

27. W. O. Kermack and A. G. McKendrick, *Proc. Royal Soc. London. Series A* **115**(772), 700 (1927).

28. C. S. Elton, *The Ecology of Invasions by Animals and Plants* (Methuen Ltd., London, 1958).

29. R. S. Steneck, M. H. Graham, B. J. Bourque et al., *Environmental Conserv.* **29**(4), 436 (2002).

30. W. N. Adger, T. P. Hughes, C. Folke et al., *Science* **309**(5737), 1036 (2005); R. Costanza, M. Daly, C. Folke et al., *Bioscience* **50**(2), 149 (2000).

31. M. Scheffer, F. Westley, and W. Brock, *Ecosystems* **6**(5), 493 (2003).

Chapter 6

1. C. H. Peterson, *Am. Nat.* **124**(1), 127 (1984).

2. J. H. Connell and W. P. Sousa, *Am. Nat.* **121**(6), 789 (1983).

3. M. Scheffer, S. R. Carpenter, J. A. Foley et al., *Nature* **413**, 591 (2001).

4. S. Rinaldi and M. Scheffer, *Ecosystems* **3**(6), 507 (2000).

5. M. Scheffer, S. Rinaldi, Y. A. Kuznetsov et al., *Oikos* **80**, 519 (1997); M. Scheffer, S. Rinaldi, and Y. A. Kuznetsov, *Can. J. Fish. Aquat. Sci.* **57**(6), 1208 (2000).

6. V. Brovkin, M. Claussen, V. Petoukhov et al., *J. Geophys. Res. Atmos.* **103**(D24), 31613 (1998); M. Claussen, C. Kubatzki, V. Brovkin et al., *Geophys. Res. Lett.* **26**(14), 2037 (1999).

7. S. H. Strogatz, *Nonlinear Dynamics and Chaos—With Applications to Physics, Biology, Chemistry, and Engineering*, 1st ed. (Addison-Wesley Publishing Company, Reading, MA, 1994).

8. C. S. Holling, *Annu. Rev. Ecol. Syst.* **4**, 1 (1973).

9. C. S. Holling, in *Engineering Resilience vs. Ecological Resilience*, P. C. Schulze, Ed., pp. 31–43 (National Academy Press, Washington DC, 1996).

10. S. Carpenter, B. Walker, J. M. Anderies et al., *Ecosystems* **4**(8), 765 (2001).

11. E. H. Van Nes and M. Scheffer, *Am. Nat.* **169**(6), 738–747 (2007).

12. B. H. Walker and D. Salt, *Resilience Thinking—Sustaining Ecosystems and People in a Changing World* (Island Press, Washington, DC, 2006).

13. S. R. Carpenter, *Regime Shifts in Lake Ecosystems: Pattern and Variation* (Ecology Institute, Oldendorf/Luhe, Germany, 2002).

Part II

Chapter 7

1. S. A. Forbes, *Bulletin of the Scientific Association (Peoria, IL)*, 77 (1887).

2. B. Moss, *Ecology of Fresh Waters, Man & Medium.*, 2nd ed. (Blackwell Scientific, Oxford, UK, 1988).

3. M. Scheffer, S. H. Hosper, M. L. Meijer et al., *Trends Ecol. Evol.* **8**(8), 275 (1993).

4. E. H. Van Nes, M. Scheffer, M. S. Van den Berg et al., *Aquat. Bot.* **72**(3–4), 275 (2002).

5. E. H. Van Nes, M. Scheffer, M. S. Van den Berg et al., *Ecol. Model.* **159**(2–3), 103 (2003).

6. J. H. Janse, E. Van Donk, and R. D. Gulati, *Neth. J. Aquat. Res.* **29**(1), 67 (1995); J. H. Janse, E. Van Donk, and T. Aldenberg, *Water. Res.* **32**(9), 2696 (1998).

7. J. H. Janse, *Hydrobiologia* **342**, 1 (1997).

8. T. L. Lauridsen, E. Jeppesen, and M. Søndergaard, *Hydrobiologia* **276**, 233 (1994); M. Scheffer, A. H. Bakema, and F. G. Wortelboer, *Aquat. Bot.* **45**(4), 341 (1993).

9. E.H.R.R. Lammens, "The Central Role of Fish in Lake Restoration and Management," in *The Ecological Bases for the Lake and Reservoir Management*, D. M. Harper, B. Brierley, A.J.D Ferguson, and G. Phillips, Eds., pp. 191–198 (Kluwer Academic Publisher, Dordrecht 1999).

10. M. Scheffer, *Ecology of Shallow Lakes*, 1st ed. (Chapman and Hall, London, 1998).

11. R. Levins, *Am. Sci.* **54**(4), 421 (1966).

12. G. J. Van Geest, F. Roozen, H. Coops et al., *Freshwater Biol.* **48**(3), 440 (2003).

13. D. A. Jackson, *Hydrobiologia* **268**(1), 9 (1993).

14. A. Hargeby, G. Andersson, I. Blindow et al., *Hydrobiologia* **280**, 83 (1994).

15. G. J. Van Geest, H. Coops, M. Scheffer et al., *Ecosystems* **10**(1), 37 (2007).

16. E. H. Van Nes, W. J. Rip, and M. Scheffer, *Ecosystems* **10**(1), 17 (2007).

17. S. F. Mitchell, D. P. Hamilton, W. S. MacGibbon et al., *Internationale Revue der Gesamten Hydrobiologie* **73**, 145 (1988); S. L. McKinnon and S. F. Mitchell, *Hydrobiologia* **279–280**, 163 (1994).

18. R. D. Gulati, E.H.R.R. Lammens, M. L. Meijer et al., *Biomanipulation Tool for Water Management. Proceedings of an International Converence Held in Amsterdam, The Netherlands, 8–11 August 1989*, 1st ed. (Kluwer Academic Publishers, Dordrecht, Boston, London, 1990).

19. S. H. Hosper and M. L. Meijer, *Ecological Engineering* **2**(1), 63 (1993).

20. R. M. Wright and V. E. Phillips, *Aquat. Bot.* **43**(1), 43 (1992).

21. N. Giles, *The Game Conservancy Ann. Rev.* **18**, 130 (1987); N. Giles, *Wildlife after Gravel: Twenty Years of Practical Research by the Game Conservancy and ARC* (Game Conservancy Ltd, Fordingbridge, Hampshire, UK, 1992).

22. R. E. Grift, A. D. Buijse, W.L.T. Van Densen et al., *Archiv für Hydrobiologie* **135**(2), 173 (2001).

23. M. Søndergaard, E. Jeppesen, and J. P. Jensen, *Archiv für Hydrobiologie* **162**(2), 143 (2005).

24. U. Sommer, Z. M. Gliwicz, W. Lampert et al., *Archiv für Hydrobiologie* **106**(4), 433 (1986).

25. H. J. Dumont, *Hydrobiologia* **272**(1–3), 27 (1994).

26. E. Jeppesen, M. Søndergaard, N. Mazzeo et al., in *Lake Restoration and Biomanipulation in Temperate Lakes: Relevance for Subtropical and Tropical Lakes,*

M. V. Reddy, Ed., pp. 331–359 (Science Publishers, Plymouth, UK, 2005); F. Scasso, N. Mazzeo, J. Gorga et al., *Aquat. Conserv.—Mar. Freshw. Ecosyst.* **11**(1), 31 (2001).

27. J. B. Grace and L. J. Tilly,. *Archiv für Hydrobiologie* **77**(4), 475 (1976); N. Rooney and J. Kalff, *Aquat. Bot.* **68**(4), 321 (2000); M. Scheffer, M. R. De Redelijkheid, and F. Noppert, *Aquat. Bot.* **42**, 199 (1992); T. A. Nelson, *Aquat. Bot.* **56**(3–4), 245 (1997).

28. G. A. Weyhenmeyer, *Ambio* **30**(8), 565 (2001); D. G. George, *Freshwater Biol.* **45**(2), 111 (2000); D. T. Monteith, C. D. Evans, and B. Reynolds, *Hydrological Processes* **14**(10), 1745 (2000); D. Straile, D. M. Livingstone, G. A. Weyhenmeyer et al., *Geophys. Monograph* **134**, 263 (2003); W. J. Rip, M. Ouboter, B. Beltman et al., *Archiv für Hydrobiologie* **164**(3), 387 (2005).

29. M. Scheffer, S. R. Carpenter, J. A. Foley et al., *Nature* **413**, 591 (2001).

30. M. Scheffer, G. J. Van Geest, K. Zimmer et al., *Oikos* **112**(1), 227 (2006).

31. E. Van Donk and R. D. Gulati, *Water Sci. Technol.* **32**(4), 197 (1995).

32. I. Blindow, *Freshwater Biol.* **28**(1), 15 (1992); I. Blindow, G. Andersson, A. Hargeby et al., *Freshwater Biol.* **30**(1), 159 (1993); I. Blindow, A. Hargeby, and G. Andersson, *Aquat. Bot.* **72**(3–4), 315 (2002); S. F. Mitchell. *Aquat. Bot.* **33**(1–2), 101 (1989).

33. M. R. Perrow, B. Moss, and J. Stansfield, *Hydrobiologia* **276**, 43 (1994); B. Moss, J. Stansfield, and K. Irvine, *Verhandlungen Internationale Vereinigung Theoretisch Angewandte Limnologie* **24**, 568 (1990).

34. J. Simons, M. Ohm, R. Daalder et al., *Hydrobiologia* **276**, 243 (1994); W. J. Rip, M.R.L. Ouboter, and H. J. Los, *Hydrobiologia* **584**(1), 415 (2007).

35. W. J. Rip, M.R.L. Ouboter, P. S. Grasshoff et al. (submitted).

36. D. L. DeAngelis, D. C. Cox, and C. C. Coutant, *Ecol. Model.* **8**, 133 (1979).

37. A. M. De Roos, L. Persson, and H. R. Thieme, *Proc. Royal Soc. London—Biol. Sci.* **270**, 611 (2003).

38. D. Claessen, A. M. De Roos, and L. Persson, *Am. Nat.* **155**(2), 219 (2000).

39. C. Luecke, M. J. Vanni, J. J. Magnuson et al., *Limnol. Oceanogr.* **35**(8), 1718 (1990); M. Boersma, O.F.R. Van Tongeren, and W. M. Mooij, *Can. J. Fish. Aquat. Sci.* **53**(1), 18 (1996).

40. M. Scheffer, S. Rinaldi, and Y. A. Kuznetsov, *Can. J. Fish. Aquat. Sci.* **57**(6), 1208 (2000).

41. M. Scheffer, *J. Plankton Res.* **13**, 1291 (1991); J. Huisman and F. J. Weissing, *Nature* **402**(6760), 407 (1999).

42. E. H. Van Nes and M. Scheffer, *Am. Nat.* **164**(2), 255 (2004).

43. J. Ringelberg, *Helgol Wiss Meeresunters* **30**(1–4), 134 (1977).

44. E. Benincà, J. Huisman, R. Heerkloss et al., *Nature* **451**(6512), 822 (2008).

45. M. S. Van den Berg, M. Scheffer, E. H. Van Nes et al., *Hydrobiologia* **409**, 335 (1999).

46. M. S. Van den Berg, H. Coops, J. Simons et al., *Aquat. Bot.* **60**(3), 241 (1998).

47. M. S. Van den Berg, "A Comparative Study of the Use of Inorganic Carbon Resources by *Chara aspera* and *Potamogeton pectinatus*," in *Charophyte Recolo-*

nization in Shallow Lakes—Processes, Ecological Effects, and Implications for Lake Management, pp. 57–67 (Vrije Universiteit Amsterdam, Amsterdam, 1999).

48. L. R. Mur, H. Schreurs, and P. Visser, "How to Control Undesirable Cyanobacterial Dominance," in *Proc. 5th International Conference on the Conservation and Management of Lakes, Stresa, Italy*, G. Giussani and C. Callieri, Eds., pp. 565–569 (International Lake Environment Committee Foundation, Otsu, Japan, 1993).

49. L. R. Mur, H. J. Gons, and L. Van Liere, *FEMS Microbiol. Lett.* 1(6), 335 (1977).

50. M. Scheffer, S. Rinaldi, A. Gragnani et al., *Ecology* 78(1), 272 (1997).

51. J. H. Janse and P.J.T.M. Van Puijenbroek, *PCDitch, een model voor eutrofiëring en vegetatie-ontwikkeling in sloten* (in Dutch, with model formulations in English), Report No. 703715 004 (RIVM, Bilthoven 1997).

52. B. Gopal, *Water Hyacinth* (Elsevier, New York, 1987); A. Mehra, M. E. Farago, D. K. Banerjee et al., *Resour. Environ. Biotechnol.* 2, 255 (1999).

53. M. Scheffer, S. Szabo, A. Gragnani et al., *Proc. Natl. Acad. Sci. USA* 100(7), 4040 (2003).

54. R. Portielje and R.M.M. Roijackers, *Aquat. Bot.* 50, 127 (1995).

55. G. E. Hutchinson, *A Treatise on Limnology. Volume III, Limnological Botany* (John Wiley and Sons, New York, 1975); P. A. Chambers, E. E. Prepas, M. L. Bothwell et al., *Can. J. Fish. Aquat. Sci.* 46, 435 (1989).

56. F. Robach, S. Merlin, T. Rolland et al., *Ecologie—Brunoy.* 27, 203 (1996); C. D. Sculthorpe, *The Biology of Aquatic Vascular Plants* (Edward Arnold Ltd, London, 1967).

57. E. Van Donk, R. D. Gulati, A. Iedema et al., *Hydrobiologia* 251, 19 (1993); R. Goulder. *Oikos* 20, 300 (1969).

58. S. R. Carpenter, *Proc. Natl. Acad. Sci. USA* 102(29), 10002 (2005).

Chapter 8

1. J. E. Lovelock and S. Epton, *New Scientist*, 304 (1975).

2. S. A. Forbes, *Bull. Scientific Association (Peoria, IL)*, 77 (1887).

3. D. C. Catling and M. W. Claire, *Earth and Planetary Sci. Lett.* 237(1–2), 1 (2005).

4. C. Goldblatt, T. M. Lenton, and A. J. Watson, *Nature* 443(7112), 683 (2006).

5. W. B. Harland and M.J.S. Rudwick, *Sci. Am.* 211(2), 28 (1964).

6. M. I. Budyko, *Tellus* 21(5), 611 (1969).

7. J. L. Kirschvink, "Late Proterozoic low-latitude global glaciation; the snowball Earth," in *The Proterozoic Biosphere: A Multidisciplinary Study*, J. W. Schopf and C. Klein, Eds., pp. 51–52 (Cambridge University Press, Cambridge, UK, 1992).

8. K. Caldeira and J. F. Kasting, *Nature* 359(6392), 226 (1992); R. T. Pierrehumbert, *Nature* 429(6992), 646 (2004).

9. R. T. Pierrehumbert, *Nature* 419(6903), 191 (2002).

10. Y. Donnadieu, Y. Godderis, G. Ramstein et al., *Nature* 428(6980), 303 (2004).

11. R. E. Kopp, J. L. Kirschvink, I. A. Hilburn et al., *Proc. Natl. Acad. Sci. USA* **102**(32), 11131 (2005).

12. P. D. Gingerich, *Trends Ecol. Evol.* **21**(5), 246 (2006).

13. L. J. Lourens, A. Sluijs, D. Kroon et al., *Nature* **435**(7045), 1083 (2005).

14. A. Tripati, J. Backman, H. Elderfield et al., *Nature* **436**(7049), 341 (2005).

15. L. R. Kump, *Nature* **436**(7049), 333 (2005).

16. J. R. Petit, J. Jouzel, D. Raynaud et al., *Nature* **399**(6735), 429 (1999).

17. A. Berger, *Rev. Geophys.* **26**(4), 624 (1988).

18. D. Paillard, *Nature* **409**(6817), 147 (2001).

19. M. Scheffer, V. Brovkin, and P. M. Cox, *Geophys. Res. Lett.* **33** (doi:10.1029/2005GL025044), L10702 (2006).

20. P. U. Clark, R. B. Alley, and D. Pollard, *Science* **286**(5442), 1104 (1999).

21. E. Rignot and P. Kanagaratnam, *Science* **311**(5763), 986 (2006).

22. H. J. Zwally, W. Abdalati, T. Herring et al., *Science* **297**(5579), 218 (2002).

23. J. A. Rial, R. A. Pielke, M. Beniston et al., *Clim. Change* **65**, 11 (2004).

24. J. Imbrie, A. Berger, E. A. Boyle et al., *Paleoceanography* **8**(6), 699 (1993).

25. M. Ghil, *Physica D* **77**(1–3), 130 (1994).

26. National Research Council, *Abrupt Climate Change: Inevitable Surprises* (U.S. National Academy of Sciences, National Research Council Committee on Abrupt Climate Change, National Academy Press, Washington, DC, 2002).

27. S. Rahmstorf, *Clim. Change* **46**(3), 247 (2000).

28. H. Stommel, *Tellus* **13**(2), 224 (1961).

29. A. Ganopolski and S. Rahmstorf, *Nature* **409**(6817), 153 (2001).

30. E. Tziperman, L. Stone, M. A. Cane et al., *Science* **264**(5155), 72 (1994).

31. M. Holmgren, M. Scheffer, E. Ezcurra et al., *Trends Ecol. Evol.* **16**(2), 89 (2001).

32. P. B. deMenocal, *Science* **292**(5517), 667 (2001).

33. F. F. Jin, J. D. Neelin, and M. Ghil, *Science* **264**(5155), 70 (1994).

34. S. D. Schubert, M. J. Suarez, P. J. Pegion et al., *Science* **303**(5665), 1855 (2004).

35. J. A. Foley, M. T. Coe, M. Scheffer et al., *Ecosystems* **6**(6), 524 (2003).

36. A. Giannini, R. Saravanan, and P. Chang, *Science* **302**(5647), 1027 (2003).

37. E. N. Lorenz, *J. Atmospheric Sci.* **20**, 130 (1963).

Chapter 9

1. S. J. Gould and N. Eldredge, *Nature* **366**(6452), 223 (1993); R. E. Lenski and M. Travisano, *Proc. Natl. Acad. Sci. USA* **91**(15), 6808 (1994); R. E. Ricklefs, *Systematic Biol.* **55**(1), 151 (2006).

2. M. Pagel, C. Venditti, and A. Meade, *Science* **314**(5796), 119 (2006).

3. D. Raup, *New Scientist* **131**(1786), 46 (1991).

4. J. W. Kirchner and A. Weil, *Nature* **404**(6774), 177 (2000).

5. D. M. Raup, *Proc. Natl. Acad. Sci. USA* **91**(15), 6758 (1994).

6. M. J. Benton, "Extinction, Biotic Replacements, and Clade Interactions," in *The Unity of Evolutionary Biology*, E. C. Dudley, Ed., pp. 89–92 (Dioscorides Press, Portland, Oregon, 1991).

7. P. Bak and K. Sneppen, *Phys. Rev. Lett.* **71**(24), 4083 (1993).

8. E. H. Van Nes and M. Scheffer, *Am. Nat.* **164**(2), 255 (2004).

9. B. J. Crespi, *Trends Ecol. Evol.* **19**(12), 627 (2004).

10. A. H. Knoll and S. B. Carroll, *Science* **284**(5423), 2129 (1999).

11. M. J. Benton and R. J. Twitchett, *Trends Ecol. Evol.* **18**(7), 358 (2003).

12. N. Lane, *Nature* **448**(7150), 122 (2007).

13. T. J. Davies, T. G. Barraclough, M. W. Chase et al., *Proc. Nat. Acad. Sci. USA* **101**(7), 1904 (2004).

14. D. I. Axelrod, *Botanical Rev.* **36**(3), 277 (1970).

15. L. J. Hickey and J. A. Doyle, *Botanical Rev.* **43**(1), 3 (1977).

16. S. L. Wing and L. D. Boucher, *Ann. Rev. Earth and Planetary Sci.* **26**, 379 (1998).

17. T. S. Feild, N. C. Arens, J. A. Doyle et al., *Paleobiology* **30**(1), 82 (2004).

18. E. H. Van Nes and M. Scheffer, *Ecology* **86**(7), 1797–1807 (2005).

19. S. C. Wang and P. Dodson, *Proc. Natl. Acad. Sci. USA* **103**(37), 13601 (2006).

20. R. A. Kerr, *Science* **302**(5649), 1314 (2003).

21. P. M. Sheehan, D. E. Fastovsky, R. G. Hoffmann et al., *Science* **254**(5033), 835 (1991).

22. R. E. Sloan, J. K. Rigby, L. M. Vanvalen et al., *Science* **232**(4750), 629 (1986).

23. P. D. Gingerich, *Trends Ecol. Evol.* **21**(5), 246 (2006).

24. M. E. Gilpin and T. J. Case, *Nature* **261**, 40 (1976).

25. J. A. Drake, *Trends Ecol. Evol.* **5**(5), 159 (1990).

26. B. Konar and J. A. Estes, *Ecology* **84**(1), 174 (2003).

27. M. Scheffer, S. H. Hosper, M. L. Meijer et al., *Trends Ecol. Evol.* **8**(8), 275 (1993).

28. T. P. Hughes, *Science* **265**, 1547 (1994).

29. M. Claussen, C. Kubatzki, V. Brovkin et al., *Geophys. Res. Lett.* **26**(14), 2037 (1999).

30. R. V. Sole, S. C. Manrubia, M. Benton et al., *Nature* **388**(6644), 764 (1997).

31. M.E.J. Newman, *J. Theor. Biol.* **189**(3), 235 (1997).

Chapter 10

1. J. H. Steele, *Fish. Res.* **25**(1), 19 (1996).

2. S. R. Hare and N. J. Mantua, *Prog. Oceanogr.* **47**(2–4), 103 (2000).

3. N. Mantua, *Prog. Oceanogr.* **60**(2–4), 165 (2004).

4. J. H. Steele, *Prog. Oceanogr.* **60**(2–4), 135 (2004).

5. W. S. Wooster and C. I. Zhang, *Prog. Oceanogr.* **60**(2–4), 183 (2004).

6. C. H. Hsieh, S. M. Glaser, A. J. Lucas et al., *Nature* **435**(7040), 336 (2005).

7. J. H. Steele and E. W. Henderson, *Philos. Trans. Royal Soc. London B Biol. Sci.* **343**, 5 (1994).

8. D. H. Cushing, *Climate and Fisheries* (Academic, London, 1983) P. 387; F. S. Russell, A. J. Southwar, G. T. Boalch et al., *Nature* **234**(5330), 468 (1971).

9. G. Beaugrand, *Prog. Oceanogr.* **60**(2–4), 245 (2004).

10. G. Beaugrand, P. C. Reid, F. Ibanez et al., *Science* **296**(5573), 1692 (2002).

11. N. Daan, *Rapp. et Proc—Verb. Cons. Int. Explor. Mer* **177**, 405 (1980); J. Alheit and M. Niquen, *Prog. Oceanogr.* **60**(2–4), 201 (2004).

12. F. P. Chavez, J. Ryan, S. E. Lluch-Cota et al., *Science* **299**(5604), 217 (2003).

13. A. Bakun and P. Cury, *Ecol. Lett.* **2**(6), 349 (1999).

14. J. H. Steele and E. W. Henderson, *Science* **224**, 985 (1984).

15. C. Walters and J. F. Kitchell, *Can. J. Fish. Aquat. Sci.* **58**(1), 39 (2001).

16. A. M. De Roos and L. Persson, *Proc. Natl. Acad. Sci. USA* **99**(20), 12907 (2002); A. M. De Roos, L. Persson, and H. R. Thieme, *Proc. Royal Soc. London—Biol. Sci.* **270**, 611 (2003).

17. J. A. Hutchings and J. D. Reynolds, *Bioscience* **54**(4), 297 (2004).

18. Millennium Ecosystem Assessment, *Ecosystems and Human Well-Being: Synthesis* (Island Press, Washington, DC., 2005).

19. I. Noy-Meir, *J. Ecol.* **63**(2), 459 (1975).

20. M. Rietkerk and J. Van de Koppel, *Oikos* **79**(1), 69 (1997); J. Van de Koppel, M. Rietkerk, and F. J. Weissing, *Trends Ecol. Evol.* **12**(9), 352 (1997).

21. J.B.C. Jackson, *Proc. Natl. Acad. Sci. USA* **98**(10), 5411 (2001).

22. G. Marteinsdottir and K. Thorarinsson, *Can. J. Fish. Aquat. Sci.* **55**(6), 1372 (1998).

23. G. A. Rose, B. deYoung, D. W. Kulka et al., *Can. J. Fish. Aquat. Sci.* **57**(3), 644 (2000).

24. K. T. Frank, B. Petrie, J. S. Choi et al., *Science* **308**(5728), 1621 (2005).

25. Q. Schiermeier, *Nature* **428**(6978), 4 (2004).

26. M. Scheffer, S. R. Carpenter, and B. De Young, *Trends Ecol. Evol.* **20**(11), 579 (2005).

27. D. M. Ware and R. E. Thomson, *Science* **308**(5726), 1280 (2005).

28. R. H. Peters, *Limnol. Oceanogr.* **31**(5), 1143 (1986).

29. C. H. Greene and A. J. Pershing, *Science* **315**(5815), 1084 (2007).

30. B. Worm and R. A. Myers, *Ecology* **84**(1), 162 (2003).

31. J.B.C. Jackson, M. X. Kirby, W. H. Berger et al., *Science* **293**(5530), 629 (2001).

32. J. C. Castilla and L. R. Duran, *Oikos* **45**(3), 391 (1985); J. C. Castilla. *Trends Ecol. Evol.* **14**(7), 280 (1999).

33. M. Nystrom, C. Folke, and F. Moberg, *Trends Ecol. Evol.* **15**(10), 413 (2000); N. Knowlton. *Am. Zool.* **32**(6), 674 (1992); T. J. Done, *Hydrobiologia* **247** (1–2), 121 (1991); L. J. McCook, *Coral Reef* **18**, 357 (1999); D. R. Bellwood, T. P. Hughes, C. Folke et al., *Nature* **429**(6994), 827 (2004).

34. T. P. Hughes, *Science* **265**, 1547 (1994).

35. T. P. Hughes, D. R. Bellwood, C. Folke et al., *Trends Ecol. Evol.* **20**(7), 380 (2005).

36. D. G. Raffaelli and S. J. Hawkins, *Intertidal Ecology* (Chapman and Hall, London, 1996).

37. R. S. Steneck, M. H. Graham, B. J. Bourque et al., *Environmental Conserv.* **29**(4), 436 (2002).

38. R. S. Steneck, J. Vavrinec, and A. V. Leland, *Ecosystems* **7**(4), 323 (2004).

39. B. Konar and J. A. Estes, *Ecology* **84**(1), 174 (2003).

40. J. A. Estes, M. T. Tinker, T. M. Williams et al., *Science* **282**(5388), 473 (1998).

41. R. E. Scheibling, A. W. Hennigar, and T. Balch, *Can. J. Fish. Aquat. Sci.* **56** (12), 2300 (1999).

42. C. B. Officer, R. B. Biggs, J. L. Taft et al., *Science* **223**(4631), 22 (1984).

43. H. S. Lenihan, F. Micheli, S. W. Shelton et al., *Limnol. Oceanogr.* **44**(3), 910 (1999).

44. L. H. Gunderson, *Ecological Economics* **37**(3), 371 (2001).

45. B. Chandler, A. M. Frank, and M. McMurry, Eds., *The New Student's Reference Work for Teachers, Students, and Families* (F. E. Compton and Company, Chicago, 1914).

46. T. Van der Heide, E. H. van Nes, G. W. Geerling et al., *Ecosystems* **10**(8), 1311 (2007).

47. M. Scheffer, R. Portielje, and L. Zambrano, *Limnol. Oceanogr.* **48**, 1920 (2003).

48. E. H. Van Nes, T. Amaro, M. Scheffer et al., *Mar. Ecol. Prog. Ser.* **330**, 39–47 (2007).

49. T. Amaro and G.C.A. Duineveld, "Benthic shift on *Amphiura filiformis* and *Calianassa subterranea* population in the period 1992–1997 at the Friesian Front (Southern North Sea)" (submitted).

50. F. Creutzberg, P. Wapenaar, G. Duineveld et al., *Rapports et proces verbaux des reunions/Commission Internationale pour l' Exploration Scientifique de la Mer Mediterranee* **183**, 101 (1984).

51. G.C.A. Duineveld and G. J. van Noort, *Neth. J. Sea Res.* **20**(1), 85 (1986).

52. T. Amaro, M. Bergman, M. Scheffer et al., *Hydrobiologia* **589**(1), 273 (2007).

53. J. Van de Koppel, P.M.J. Herman, P. Thoolen et al., *Ecology* **82**(12), 3449 (2001).

Chapter 11

1. M. Scheffer, M. Holmgren, V. Brovkin et al., *Global Change Biol.* **11**(7), 1003 (2005).

2. J. G. Charney, *J. Royal Meteorological Soc.* **101**, 193 (1975).

3. Y. K. Xue and J. Shukla, *J. Climate* **6**(12), 2232 (1993).

4. N. Zeng, J. D. Neelin, K. M. Lau et al., *Science* **286**(5444), 1537 (1999).

5. A. Kleidon and M. Heimann, *Clim. Dyn.* **16**(2–3), 183 (2000).

6. A. J. Dolman, M.A.S. Dias, J. C. Calvet et al., *Ann. Geophys.—Atmos. Hydrospheres Space Sci.* **17**(8), 1095 (1999).

7. V. Brovkin, M. Claussen, V. Petoukhov et al., *J. Geophys. Res. Atmos.* **103** (D24), 31613 (1998).

8. L. D. Sternberg, *Global Ecol. Biogeogr.* **10**(4), 369 (2001).

9. J. A. Foley, M. T. Coe, M. Scheffer et al., *Ecosystems* **6**(6), 524 (2003).

10. M. Claussen, *Clim. Dyn.* **13**(4), 247 (1997).

11. P. Hoelzmann, D. Jolly, S. P. Harrison et al., *Global Biogeochem. Cycles* **12**(1), 35 (1998); D. Jolly, I. C. Prentice, R. Bonnefille et al., *J. Biogeogr.* **25**(6), 1007 (1998).

12. P. deMenocal, J. Ortiz, T. Guilderson et al., *Quaternary Sci. Rev.* **19**(1–5), 347 (2000).

13. M. Claussen, C. Kubatzki, V. Brovkin et al., *Geophys. Res. Lett.* **26**(14), 2037 (1999).

14. M. D. Oyama and C. A. Nobre, *Geophys. Res. Lett.* **30**(23) (2003).

15. E. R. Fuentes, R. D. Otaiza, M. C. Alliende et al., *Oecologia* **62**, 405 (1984).

16. H. T. Dublin, A. R. Sinclair, and J. McGlade, *J. Anim. Ecol.* **59**(3), 1147 (1990).

17. A. Dobson and W. Crawley, *Trends Ecol. Evol.* **9**(10), 393 (1994).

18. M. Holmgren, B. C. Lopez, J. R. Gutierrez et al., *Global Change Biol.* **12**(12), 2263 (2006).

19. M. A. Bravo-Ferro and M. Rodriguez-Sánchez, *Zonas Aridas* **7**, 206 (2003).

20. M. Rietkerk and J. Van de Koppel, *Oikos* **79**(1), 69 (1997).

21. J. Van de Koppel, M. Rietkerk, and F. J. Weissing, *Trends Ecol. Evol.* **12**(9), 352 (1997); M. Shachak, M. Sachs, and I. Moshe, *Ecosystems* **1**(5), 475 (1998).

22. M. Holmgren, M. Scheffer, and M. A. Huston, *Ecology* **78**(7), 1966 (1997).

23. J. B. Wilson and A.D.Q. Agnew, *Adv. Ecol. Res.* **23**, 263 (1992).

24. R. Geiger, *The Climate Near the Ground* (Harvard University Press, Cambridge, MA, 1965); R. Joffre and S. Rambal, *Acta Oecologica—Oecologia Plantarum* **9**(4), 405 (1988).

25. A. Valientebanuet and E. Ezcurra, *J. Ecol.* **79**(4), 961 (1991).

26. R. M. Callaway, *Botan. Rev.* **61**(4), 306 (1995).

27. M. Holmgren and M. Scheffer, *Ecosystems* **4**(2), 151 (2001).

28. J. von Hardenberg, E. Meron, M. Shachak et al., *Phys. Rev. Lett.* **8719**(19), art. no. 198101 (2001); R. HilleRisLambers, M. Rietkerk, F. Van den Bosch et al., *Ecology* **82**(1), 50 (2001); J. Van de Koppel and M. Rietkerk, *Am. Nat.* **163**(1), 113 (2004); M. Rietkerk, M. C. Boerlijst, F. van Langevelde et al., *Am. Nat.* **160**(4), 524 (2002); M. Rietkerk, S. C. Dekker, P. C. de Ruiter et al., *Science* **305**(5692), 1926 (2004).

29. M. Y. Bader, *Tropical Alpine Treelines, How Ecological Processes Control Vegetation Patterning and Dynamics* (Wageningen University, Wageningen, The Netherlands, 2007).

30. H. F. Sklar and G. A. van der Valk, Eds., *Tree Islands of the Everglades* (Kluwer Academic Publishers, Dordrecht, The Netherlands, 2002).

31. G. B. Bonan, D. Pollard, and S. L. Thompson, *Nature* **359**(6397), 716 (1992).

32. V. Brovkin, S. Levis, M. F. Loutre et al., *Clim. Change.* **57**(1–2), 119 (2003).

33. J.P.P. Jasinski and S. Payette, *Ecological Monographs* **75**(4), 561 (2005).

34. R. T. Paine, M. J. Tegner, and E. A. Johnson, *Ecosystems* **1**, 535 (1998).

35. C. S. Holling, "The Spruce-Budworm/Forest-Management Problem," in *Adaptive Environmental Assessment and Management,* C. S. Holling, Ed., pp. 143–182 (John Wiley & Sons, New York, 1978).

36. D. Ludwig, D. D. Jones, and C. S. Holling, *J. Anim. Ecol.* **47**(1), 315 (1978).

37. N. Van Breemen, *Trends Ecol. Evol.* **10**(7), 270 (1995).

38. L.P.M. Lamers, R. Bobbink, and J.G.M. Roelofs, *Global Change Biology* **6**(5), 583 (2000).

39. I. Hanski, *Nature* **396**(6706), 41 (1998).

40. I. Hanski and M. Gyllenberg, *Am. Nat.* **142**(1), 17 (1993).

41. I. Hanski, J. Poyry, T. Pakkala et al., *Nature* **377**, 618 (1995).

42. M. Loreau, N. Mouquet, and R. D. Holt, *Ecol. Lett.* **6**(8), 673 (2003).

43. M. Scheffer, G. J. van Geest, K. Zimmer et al., *Oikos* **112**(1), 227 (2006).

44. W. O. Kermack and A. G. McKendrick, *Proc. Royal Soc. London Series A* **115**(772), 700 (1927).

45. D.J.D. Earn, P. Rohani, B. M. Bolker et al., *Science* **287**(5453), 667 (2000).

46. F. Berendse and W. T. Elberse, "Competition and Nutrient Availablity in Heathlands and Grassland Ecosystems," in *Perspectives on Plant Competition,* J. B. Grace and D. Tilman, Eds., pp. 94–116 (Academic Press, New York, 1990).

Chapter 12

1. P. B. deMenocal, *Science* **292**(5517), 667 (2001).

2. J. Diamond, *Collapse: How Societies Choose to Fail or Survive* (Viking Adult, 2004).

3. S. Rinaldi, *Applied Mathematics and Computation* **95**(2–3), 181 (1998).

4. S. Rinaldi, G. Feichtinger, and F. Wirl, *Complexity* **3**(5), 53 (1998).

5. P. T. Coleman, R. Vallacher, A. Nowak et al., *American Behavioral Scientist* **50**(11), 1454 (2007).

6. J. Moffat, *Complexity Theory and Network-Centric Warfare* (DOD Command and Control Research Program, Washington, DC, 2003).

7. M. Scheffer and F. R. Westley, *Ecology and Society* **12**(2), 36 (2007).

8. Y. C. Wang and E. L. Ferguson, *Nature* **434**(7030), 229 (2005).

9. C. P. Bagowski and J. E. Ferrell, *Current Biology* **11**(15), 1176 (2001).

10. T. C. Chamberlin, *J. Geol.* **5**, 837 (1897).

11. M. Gladwell, *Blink: The Power of Thinking without Thinking* (Little, Brown, and Co., New York, 2005).

12. G. Klein, *Sources of Power* (MIT Press, Boston, 1999).

13. M. R. Kruk, J. Halasz, W. Meelis et al., *Behavioral Neuroscience* **118**(5), 1062 (2004).

14. H. R. Arkes and P. Ayton, *Psychol. Bull.* **125**(5), 591 (1999).

15. R. Dawkins and T. R. Carlisle, *Nature* **262**(5564), 131 (1976).

16. B. M. Staw and H. Hoang, *Admin. Sci. Quarterly* **40**(3), 474 (1995).

17. H. R. Arkes, *J. Behavioral Decision Making* **9**(3), 213 (1996).

18. J. Brockner, *Academy of Management Rev.* **17**(1), 39 (1992).

19. L. P. Gerlach and V. H. Hine, *People, Power, Change: Movements of Social Transformation* (Bobbs-Merrill, Indianapolis, 1970).

20. J. Darley and B. Latane, *J. Personality and Social Psychol.* **8**, 377 (1968).

21. L. Festinger, H. W. Riecken, and S. Schachter, *When Prophecy Fails* (Harper and Row, New York, 1956).

22. S. Asch, *Sci. Am.* **193**, 31 (1955).

23. S. Milgram, *Psychology Today* **1**, 60 (1967).

24. E. M. Rogers, *Diffusion of Innovations* (Free Press, New York, 1983).

25. E. Hatfield, J. Copioppo, and R. Rapson, *Emotional Contagion* (Cambridge University Press, Cambridge, UK, 1994).

26. T. Kuhn, *The Structure of Scientific Revolution* (University of Chicago, Chicago, 1962).

27. W. Brock and S. Durlauf, *Econ. Theory* **14**, 113 (1999).

28. J. A. Holyst, K. Kacperski, and F. Schweitzer, *Ann. Rev. Computational Phys.* **9**, 253 (2002); K. Kacperski and J. A. Holyst. *Physica A* **269**(2–4), 511 (1999).

29. G. Schwartz and J. J. Nicols, *After Collapse: The Regeneration of Complex Societies* (The University of Arizona Press, Tucson, 2006).

30. M. A. Janssen, T. A. Kohler, and M. Scheffer, *Current Anthropology* **44**(5), 722 (2003).

31. E. Boulding, "Power and Conflict in Organizations: Further Reflections on Conflict Management," in *Power and Conflict in Organizations*, R. L. Kahn and E. Boulding, Eds., pp. 146–150 (Basic Books, New York, 1964).

32. A. Sih, A. Bell, and J. C. Johnson, *Trends Ecol. Evol.* **19**(7), 372 (2004); J. M. Koolhaas, S. M. Korte, S. F. De Boer et al., *Neurosci. Biobehav. Rev.* **23**(7), 925 (1999).

33. D. A. Levinthal and J. G. March, *Strategic Management J.* **14**, 95 (1993); J. March, *Primer on Decision Making: How Decisions Happen* (Free Press, New York, 1994).

34. C. Perrow, *Organizational Dynamics* **2**(1), 2 (1973).

35. B. Quinn, *Harvard Bus. Rev.* **May–June**, 73 (1985); J. T. Kidder, *The Soul of a New Machine* (Little, Brown, Boston, 1981); R. M. Kanter, *The Change Masters* (Free Press, New York, 1985).

36. H. Mintzberg and F. Westley, *Strategic Management J.* **13**, 39 (1992).

37. I. L. Janis, *Victims of Groupthink: A Psychological Study of Foreign-Policy Decisions and Fiascoes* (Houghton-Mifflin, New York, 1972).

38. M. Scheffer, F. Westley, and W. Brock, *Ecosystems* **6**(5), 493 (2003).

39. R. Inglehart and W. E. Baker, *Amer. Sociol. Rev.* **65**(1), 19 (2000).

Part III

Chapter 14

1. M. Scheffer and S. R. Carpenter, *Trends Ecol. Evol.* **18**(12), 648 (2003).

2. G.E.P. Box, G. M. Jenkins, and G. C. Reinsel, *Time Series Analysis: Forecasting and Control* (Prentice-Hall, Englewood Cliffs, NJ, 1994); A. R. Ives, B. Dennis, K. L. Cottingham et al., *Ecol. Monographs* **73**(2), 301 (2003); S. R. Hare and N. J. Mantua, *Prog. Oceanogr.* **47**(2–4), 103 (2000).

3. W. A. Brock, W. D. Dechert, B. LeBaron et al., *Economic Rev.* **15** (197–235) (1996).

4. S. R. Carpenter and M. L. Pace, *Oikos* **78**(1), 3 (1997).

5. M. Liermann and R. Hilborn, *Can. J. Fish. Aquat. Sci.* **54**(9), 1976 (1997).

6. S. R. Carpenter, "Alternate States of Ecosystems: Evidence and Some Implications," in *Ecology: Achievement and Challenge*, M .C. Press, N. Huntly, and S. Levin, Eds., pp. 357–381 (Blackwell, London, 2001).

7. J. P. Sutherland, *Am. Nat.* **108**(964), 859 (1974).

8. J. H. Connell and W. P. Sousa, *Am. Nat.* **121**(6), 789 (1983).

9. M. Holmgren, M. Scheffer, E. Ezcurra et al., *Trends Ecol. Evol.* **16**(2), 89 (2001).

10. C. M. Taylor and A. Hastings, *Ecol. Lett.* **8**(8), 895 (2005).

11. B. Konar and J. A. Estes, *Ecology* **84**(1), 174 (2003).

12. G. J. Van Geest, F. Roozen, H. Coops et al., *Freshwater Biol.* **48**(3), 440 (2003).

13. B. Efron and R. J. Tibshirani, *An Introduction to the Bootstrap* (Chapman and Hall, New York, 1993).

14. T. D. Havlicek and S. R. Carpenter, *Limnol. Oceanogr.* **46**(5), 1021 (2001).

15. M. Scheffer, S. Szabo, A. Gragnani et al., *Proc. Natl. Acad. Sci. USA* **100**(7), 4040 (2003).

16. M. Scheffer, S. Rinaldi, A. Gragnani et al., *Ecology* **78**(1), 272 (1997).

17. R. Hilborn and M. Mangel, *The Ecological Detective* (Princeton University Press, Princeton, NJ, 1993).

18. N. Giles, *Wildlife after Gravel: Twenty Years of Practical Research by the Game Conservancy and ARC* (Game Conservancy Ltd, Fordingbridge, Hampshire, UK, 1992).

19. J. A. Drake, G. R. Huxel, and C. L. Hewitt, *Ecology* **77**, 670 (1996).

20. M. L. Meijer, E. Jeppesen, E. Van Donk et al., *Hydrobiologia* **276**, 457 (1994).

21. M. Scheffer, *Ecology of Shallow Lakes*, 1st ed. (Chapman and Hall, London, 1998).

22. R. A. Matthews, W. G. Landis, and G. B. Matthews, *Environ. Toxicol. Chem.* **15**, 597 (1996).

23. E. Van Donk and R. D. Gulati, *Water Sci. Technol.* **32**(4), 197 (1995).

24. T. M. Frost, S. R. Carpenter, A. R. Ives et al., "Species Compensation and

Complementarity in Ecosystem Function," in *Linking Species and Ecosystems*, C. Jones and J. Lawton, Eds., pp. 224–239 (Chapman and Hall, New York, 1995).

25. M. L. Meijer, "Biomanipulation in the Netherlands—15 Years of Experience," PhD Thesis (Wageningen University, Wageningen, The Netherlands, 2000) p. 208; S. R. Carpenter, D. Ludwig, and W. A. Brock, *Ecol. Appl.* **9**(3), 751 (1999).

26. D. J. Augustine, L. E. Frelich, and P. A. Jordan, *Ecol. Appl.* **8**(4), 1260 (1998).

27. J. B. Wilson and A.D.Q. Agnew, *Adv.Ecol.Res.* **23**, 263 (1992).

28. D. A. Lashof, B. J. DeAngelo, S. R. Saleska et al., *Annu. Rev. Energ. Environ.* **22**, 75 (1997); D. A. Lashof., *Clim. Change.* **14**(3), 213 (1989); W. W. Kellogg, *J. Geophys. Res.* **88C**, 1263 (1983).

29. P. M. Cox, R. A. Betts, C. D. Jones et al., *Nature* **408**(6813), 750 (2000); P. Friedlingstein, L. Bopp, P. Ciais et al., *Geophys. Res. Lett.* **28**(8), 1543 (2001).

30. M. Scheffer, V. Brovkin, and P. M. Cox, *Geophys. Res. Lett.* **33** (doi:10.1029/2005GL025044), L10702 (2006).

31. M. Scheffer and J. Beets, *Hydrobiologia* **276**, 115 (1994).

32. T. C. Chamberlin, *J. Geol.* **5**, 837 (1897).

33. J. F. Quinn and A. E. Dunham, *Am. Nat.* **122**(5), 602 (1983); J. Roughgarden. *Am. Nat.* **122**, 583 (1983).

34. F. P. Chavez, J. Ryan, S. E. Lluch-Cota et al., *Science* **299**(5604), 217 (2003).

35. Q. Schiermeier, *Nature* **428**(6978), 4 (2004).

36. S. R. Carpenter, *Ecology* **77**(3), 677 (1996).

37. J. A. Bloomfield, R. A. Park, D. Scavia et al., in *Aquatic Modeling in the Eastern Deciduous Forest Biome, U.S.— International Biological Program*, E. Middlebrooks, D. H. Falkenberg, and T. E. Maloney, Eds., pp. 139 (Ann Arbor Science, Ann Arbor, MI, 1974).

38. N. Oreskes, K. Shraderfrechette, and K. Belitz, *Science* **263**(5147), 641 (1994).

39. R. Levins, *Am. Sci.* **54**(4), 421 (1966).

40. S. Rahmstorf, *Nature* **379**(6568), 847 (1996).

41. D. Ludwig, W. A. Brock, and S. R. Carpenter, *Ecol. Soc.* **10**(2) (2005).

Chapter 15

1. M. Scheffer, *Ecology of Shallow Lakes*, 1st ed. (Chapman and Hall, London, 1998).

2. E. H. Van Nes and M. Scheffer, *Am. Nat.* **169**(6), 738–747 (2007).

3. O. Ovaskainen and I. Hanski, *Theor. Popul. Biol.* **61**(3), 285 (2002).

4. A. R. Ives, *Ecol. Monographs* **65**, 217 (1995).

5. V. Dakos, M. Scheffer, E. H. van Nes et al., *Proc. Natl. Acad. Sci. USA* **105**(38, 14308 (2008).

6. S. R. Carpenter and W. A. Brock, *Ecol. Lett.* **9**(3), 308 (2006).

7. A. Ganopolski and S. Rahmstorf. *Nature* **409**(6817), 153 (2001).

8. R. V. Solé, S. C. Manrubia, B. Luque et al., *Complexity* **1**(5), 13 (1996).

9. S. Kefi, M. Rietkerk, C. L. Alados et al., *Nature* **449**(7159), 213 (2007).

10. M. Pascual and F. Guichard, *Trends in Ecology and Evolution* **20**(2), 88

(2005).

   11. P. E. McSharry, L. A. Smith, and L. Tarassenko, *Nat. Med.* **9**(3), 241 (2003).

   12. C. E. Elger and K. Lehnertz, *Eur. J. Neurosci.* **10**(2), 786 (1998).

   13. B. Litt, R. Esteller, J. Echauz et al., *Neuron* **30**(1), 51 (2001).

   14. J. E. Skinner, C. M. Pratt, and T. Vybiral, *Am. Heart J.* **125**(3), 731 (1993).

Chapter 16

   1. M. Scheffer, W. Brock, and F. Westley, *Ecosystems* **3**(5), 451 (2000).

   2. E. H. Van Nes, M. S. Van den Berg, J. S. Clayton et al., *Hydrobiologia* **415**, 335 (1999).

   3. D. Ludwig, S. R. Carpenter, and W. A. Brock, *Ecol. Appl.* **13**(4), 1135 (2003); S. R. Carpenter, D. Ludwig, and W. A. Brock, *Ecol. Appl.* **9**(3), 751 (1999).

   4. G. Hardin, *Science* **131**, 1292 (1960).

   5. A. Smith, *The Wealth of Nations* (Modern Library, New York, 1937).

   6. T. Dietz, E. Ostrom, and P. C. Stern, *Science* **302**(5652), 1907 (2003).

   7. D. McCloskey, *The Applied Theory of Price* (Macmillan, New York, 1982).

   8. S. Magee, W. Brock, and L. Young, *Black Hole Tariffs and Endogenous Policy Theory: Political Economy in General Equilibrium* (Cambridge University Press, Cambridge, UK, 1989).

   9. E. Ostrom, G. Gardner, and J. Walker, *Rules, Games, and Common Pool Resources* (University of Michigan Press, Ann Arbor, 1994).

   10. M. Scheffer, F. Westley, and W. Brock, *Ecosystems* **6**(5), 493 (2003).

   11. G. Klein, *Sources of Power* (MIT Press, Cambridge, MA, 1998).

   12. T. Colborn, J. Peterson Myers, and D. Dumanoski, *Our Stolen Future* (Little, Brown & Co., Boston, 1996).

   13. M. R. Taylor, P. Holmes, R. Duarte Davidson et al., *Sci. Total Environ.* **233**(1–3), 181 (1999).

   14. H. Mintzberg, *Power In and Around Organizations* (Prentice Hall, New York, 1983).

   15. R. Pascale, *The Art of Japanese Management* (Simon and Schuster, New York, 1981).

Chapter 17

   1. J. Hrbacek, M. Dvorakova, V. Korinek et al., *Verhandlungen Internationale Vereinigung Theoretisch Angewandte Limnologie* **14**, 192 (1961).

   2. J. L. Brooks and S. I. Dodson, *Science* **150**, 28 (1965).

   3. E. Van Donk, M. P. Grimm, R. D. Gulati et al., *Hydrobiologia* **200–201**, 275 (1990).

   4. S. H. Hosper and M. L. Meijer, *Ecol. Eng.* **2**(1), 63 (1993).

   5. B. Moss, J. Madgewick, and G. Phillips, *A Guide to the Restoration of Nutrient-Enriched Shallow Lakes* (Broads Authority/Environment Agency, Norwich, Norfolk, UK, 1996).

   6. T. L. Lauridsen, J. P. Jensen, E. Jeppesen et al., *Hydrobiologia* **506**(1–3), 641

(2003); M. L. Meijer, I. De Boois, M. Scheffer et al., *Hydrobiologia* **408/409**, 13 (1999).

7. M. Holmgren and M. Scheffer, *Ecosystems* **4**(2), 151 (2001); M. Holmgren, B. C. López, J. R. Gutiérrez et al., *Global Change Biology* **12**, 2263 (2006).

8. R. Adler, *Nature* **414**(6863), 480 (2001).

9. S. R. Carpenter, D. Ludwig, and W. A. Brock, *Ecol. Appl.* **9**(3), 751 (1999).

10. B. Ibelings, R. Portielje, E. Lammens et al., *Ecosystems* **10**(1), 4 (2007).

11. R. Portielje and R. E. Rijsdijk, *Freshwater Biol.* **48**(4), 741 (2003).

12. T. P. Hughes, L. H. Gunderson, C. Folke et al., *Ambio* **36**(7), 586 (2007).

13. T. P. Hughes, A. H. Baird, D. R. Bellwood et al., *Science* **301**(5635), 929 (2003).

14. C. Walters and J. F. Kitchell, *Can. J. Fish. Aquat. Sci.* **58**(1), 39 (2001); J. A. Hutchings and J. D. Reynolds, *Bioscience* **54**(4), 297 (2004).

15. F. Berkes, T. P. Hughes, R. S. Steneck et al., *Science* **311**(5767), 1557 (2006).

16. S. Rahmstorf, *Clim. Change.* **46**(3), 247 (2000).

17. S. M. Howden, S. Crimp, J. Carter et al., *Enhancing Natural Resource Management by Incorporating Climate Variability into Tree Establishment Decisions*, Final Report for the Australian Greenhouse Office (Canberra, 2004).

Appendix

1. J. Bascompte, *Annales Zoologici Fennici* **40**(2), 99 (2003).

2. C. S. Holling, *The Canadian Entomologist* **91**(7), 385 (1959).

3. M. Scheffer, V. Brovkin, and P. M. Cox, *Geophys. Res. Lett.* **33** (doi:10.1029/2005GL025044), L10702 (2006).

4. E. H. Van Nes and M. Scheffer, *Am. Nat.* **164**(2), 255 (2004).

5. M. Scheffer and S. Rinaldi, *Freshwater Biol.* **45**(2), 265 (2000).

6. M. Scheffer and R. J. De Boer, *Ecology* **76**(7), 2270 (1995).

7. M. Scheffer, S. Rinaldi, Y. A. Kuznetsov et al., *Oikos* **80**, 519 (1997).

8. J. von Hardenberg, E. Meron, M. Shachak et al., *Phys. Rev. Lett.* **8719**(19), Art. No. 198101 (2001); M. Rietkerk, S. C. Dekker, P. C. de Ruiter et al., *Science* **305**(5692), 1926 (2004).

9. M. Rietkerk, M. C. Boerlijst, F. van Langevelde et al., *Am. Nat.* **160**(4), 524 (2002).

10. M. Scheffer, *Ecology of Shallow Lakes*, 1st ed. (Chapman and Hall, London, 1998).

11. M. Scheffer, S. Szabo, A. Gragnani et al., *Proc. Natl. Acad. Sci. USA* **100**(7), 4040 (2003).

12. J. Andreoni, *J. Political Economy* **106**(6), 1186 (1998).

13. W. Brock and S. Durlauf, *Econ. Theory* **14**, 113 (1999); M. Scheffer, F. Westley, and W. Brock, *Ecosystems* **6**(5), 493 (2003).

Index

Milton Keynes UK
Ingram Content Group UK Ltd.
UKHW021809241123
433186UK00010B/372